D1767384

Building Student Safety Habits for the Workplace

Instructor's Edition

Photo Credits

On the cover: 4-L bottle of acetone and 500-mL beaker, photos courtesy of EM Science, a division of EM Industries, Inc.; eyewash station and drench shower, picture courtesy of Haws Corporation; and squeeze bottle, photo courtesy of Nalge Nunc International.

PACT Chemical Technology Resource Series

Building Student Safety Habits for the Workplace

Series Editor

Mickey Sarquis
> Director, Center for Chemical Education, Miami University

Contributing Authors

Susan Gertz
> Manager, Terrific Science Press, Miami University

James Kaufman
> Director, Laboratory Safety Institute

Mark W. Meszaros
> Vice President, Technical Services, Flinn Scientific, Inc.

Brian Shmaefsky
> Director of Biotechnology Education, Kingwood College

Mary Ann Solstad
> Solstad Health & Safety Evaluations

C. Michael Woodward
> Hazardous Residuals Manager/Laboratory Safety Coordinator, Miami University

Terrific Science Press
Miami University Middletown
Middletown, Ohio

Terrific Science Press
Miami University Middletown
4200 E. University Blvd.
Middletown, Ohio 45042
513/727-3269
cce@muohio.edu
© 2000 by Terrific Science Press

This book is intended to provide guidance for student activities in chemical technology programs. The contents of this book are intended to be informational only and are not designed to apply to specific safety situations. Where student activities include laboratory work, these laboratory activities will have unique hazards and will require appropriate policies, procedures, controls, and management. The guidance provided in this publication and many of the references may need to be modified to fit a particular laboratory and its unique activities, equipment, supplies, and location. Federal, state, local, or institutional standards, codes, and regulations may also apply and supercede information found in this guide or its references. This publication does not substitute for review of applicable government regulations and standards. This book is based on current standards and regulations as of the date of publication. The user should check existing regulations as they are updated. No warranty, guarantee, or representation is made by the authors or the Terrific Science Press as to the correctness or sufficiency of any information herein. Neither the authors nor the publisher assumes any responsibility or liability for the use of the information herein, nor can it be assumed that all necessary warnings and precautionary measures are contained in this publication. Other or additional information or measures may be required or desirable because of particular or exceptional conditions or circumstances, or because of new or changed legislation.

ISBN: 1-883822-18-1

Printed in the United States of America. All rights reserved. The publisher takes no responsibility for the use of any materials or methods described in this book, nor for the products thereof. Permission is granted to copy materials for classroom use.

This material is based upon work supported by the National Science Foundation under Grant Number DUE-9454518. Any opinions, findings, and conclusions or recommendations expressed in this material are those of the authors and do not necessarily reflect the views of the National Science Foundation.

Table of Contents

Acknowledgments

The authors and editor wish to thank the following individuals who contributed to the development of *Building Student Safety Habits for the Workplace:*

Terrific Science Press Design and Production Team

Document Production Manager: Susan Gertz
Technical Editing: Lisa Taylor, Jenny Stencil, Kim Jacobs
Illustrations: Carole Katz, Lisa Taylor
Technical Writing: Tom Schaffner, Lisa Taylor, Jenny Stencil, Christine Mulvin, Becky Franklin, Kim Jacobs, Lisa Alexander
Design/Layout: Susan Gertz, Becky Franklin
Production: Lisa Taylor, Jenny Stencil, Tom Schaffner, James Malayang, Lisa Alexander, Christine Mulvin, Becky Franklin, Brian Fair, Kim Jacobs, Jan Garcia
Cover Design: Susan Gertz

Special thanks to the following for providing content in their specialties

Charles Beel—Fire Marshal, Middletown Fire Department, retired
John Bronaugh—Haws Corporation
Blair Brewster—Electromark USA
Dave Erisman—Quantum Chemical Corporation, retired
Scott Geller—Center for Applied Behavior Systems
Don Groce—Best Manufacturing Company
Lynn Hogue—Associate Director, Center for Chemical Education, Miami University Middletown
Neal Langerman—Advanced Chemical Safety
Baird Lloyd—Center for Chemical Education, Miami University Middletown
Tony MacErlane—MG Industries
Carl Morgan—Department of Engineering Technology, retired, and Center for Chemical Education, Miami University Middletown
Joe Partlow—Department of Chemistry and Biochemistry, Miami University
Phil Redmond—formerly of Quantum Chemical Corporation
Laura Rosato—formerly of Quantum Chemical Corporation
Mark Sabo—Department of Chemistry, Catawba College
C. Nelson Schlatter—Ansell Protective Products
Jeffrey O. Stull—President, International Personnel Protection, Inc.
Jackie Webster—Chemical Laboratory Coordinator, Miami University Middletown
Matt Williams—Quantum Chemical Corporation, retired
Jim Zeigler—DuPont Tyvek® and Tychem® Protective Apparel

Special thanks to the following for providing chemical labels, cover photographs, and safety cartoons

John Elfers—Fisher Scientific
Ted Goff—Business and Safety Cartoons, *www.tedgoff.com/*
Dan Haggerty—Fisher Scientific
Rande Kline—EM Science
Rebecca Vaiarelli—EM Science
P. Joseph Yachanin—Luttner & Yachanin Advertising, Inc.

Reviewers and Testers

W.H. "Jack" Breazeale, Jr.—College of Charleston; Former Chair, American Chemical Society Committee on Chemical Safety

John Bronaugh—Haws Corporation (Chapter 7)

Eric Buck—Center for Chemical Education, Miami University Middletown

Ken Chapman—Head, Technician Resources/Education, American Chemical Society, retired

David Michael Coons—Director, Radiation and Safety Officer, Environmental Health and Safety, Miami University

Hank Greeb—Chemical Engineer, The Procter & Gamble Company, retired

Rudy Gerlach—Gerlach Training and Consulting

Lynn Hogue—Associate Director, Center for Chemical Education, Miami University Middletown

Amy Hudepohl—Center for Chemical Education, Miami University Middletown

John Kenkel—Environmental Laboratory Technology, Southeast Community College

Neal Langerman—Advanced Chemical Safety (Chapter 2)

James Laughlin—Department of Chemistry and Biochemistry, Miami University Middletown

Tina Leaym—Chemist, Dow Corning Corporation

Ethel McLaughlin—Research Technician, The Dow Chemical Company

Carl Morgan—Professor, Department of Engineering Technology, retired, and Center for Chemical Education, Miami University Middletown (Chapter 5)

Connie Murphy—Senior Research Technologist, The Dow Chemical Company

Ian Peat—Department of Chemistry and Biochemistry, Miami University

Jerry Sarquis—Professor, Department of Chemistry and Biochemistry, Miami University

Clifford L. Schrader—Program Manager, Summit County Public Schools, retired

Amy Stander—Assistant Director, Partnership for the Advancement of Chemical Technology, Center for Chemical Education, Miami University Middletown

Linda Woodward—Center for Chemical Education, Miami University Middletown

Foreword

From their first day of employment, laboratory staff working in the chemical processes and allied industries are expected to understand and practice safe laboratory behaviors. Educators in chemistry, chemical technology, and related programs have a responsibility to make safety a significant and meaningful part of their students' education.

Because safety education is such an essential part of education, we have developed *Building Student Safety Habits for the Workplace* as a resource and activity book for chemistry and chemical technology faculty and students. *Building Student Safety Habits for the Workplace* is not intended to duplicate or replace the many excellent and comprehensive reference books on chemical safety—notably *Prudent Practices in the Laboratory: Handling and Disposal of Chemicals* from the National Research Council. Rather, this PACT Chemical Technology Resource series book is intended to serve as a companion instructional guide that busy instructors can use as a convenient starting point for building students' safety knowledge and skills. The exercises and labs in this book can be used in sequence or in parts as needed. We suggest that you use these materials as a component of all regular laboratory classes rather than isolate them in a stand-alone course; we believe that a true safety culture can be attained only if students routinely consider safety as an integral part of everything they do.

This volume was created through a major collaborative effort that included safety experts in both industry and academia, chemists and chemical technicians, materials developers, and technical writers. All who have contributed to the development, laboratory testing, and classroom testing of this document have added significantly to its usefulness as an instructional resource. We thank the entire team for their dedication to the project.

The PACT Chemical Technology Resource series is a product of the National Science Foundation-funded Partnership for the Advancement of Chemical Technology (PACT) program at Miami University Middletown in Ohio. PACT is an industrial/academic collaborative committed to creating a well-educated, chemistry-based technical workforce. Members of the PACT Consortium share the goal of bringing chemistry and chemical technology education into closer alignment with the skills, methods, problem solving, and content used in today's industrial and government laboratories. We invite you to learn more about PACT by visiting our website at *www.terrificscience.org/PACT/*.

Mickey Sarquis, Director
Center for Chemical Education

Introduction

Building Student Safety Habits for the Workplace comprises a student edition and an instructor's edition (this book). The student edition (see cover on left) provides background readings along with instructions for exercises and labs that are designed to engage students in applying the safety information they learn. (The student edition is available from Terrific Science Books, Kits, and More, 517/727-3269, *www.terrificscience.org*. ISBN: 1-883822-18-1.) The background information can be extended by having students read other safety resources. A list of highly recommended resources is included in Chapter 1 in the section "Developing Your Safety Knowledge and Skills." We suggest that students begin collecting the resources from this list to build their own safety library for use throughout their professional careers.

The instructor's edition contains the full content of the student book (these pages are marked "student pages" with a gray bar on the outside of the page), along with teaching ideas, suggested demonstrations, setup notes for labs and demonstrations, and sample student data for labs. You will note that where appropriate, student answers are provided to questions that are not subjective or dependent on individual circumstances. (Of course, these answers do not appear in the student edition of the book.) Also note that figures in the student pages are numbered sequentially in each chapter, while diagrams in the instructor notes are not numbered.

Safety Culture

The theme of safety culture is central to *Building Student Safety Habits for the Workplace*. Chapter 1 introduces this idea to students; and the readings, demonstrations, paper-and-pencil exercises, and labs in the subsequent chapters of the book all reinforce this concept. To contribute to students' preparation for the professional chemical workplace and to protect them as they learn, we encourage you to reinforce a safety culture as you use the materials in this book and to explore other creative ways to integrate safety-oriented attitudes and practices throughout your classes.

One way to cultivate a safety culture is to conduct safety meetings with students to model the safety meetings they are likely to encounter on the job and to reinforce the safety ideas being taught concurrently in the laboratory. Students can take turns being responsible for safety meetings by leading discussions and selecting topics. Additionally, you may wish to have students design and implement laboratory safety audits as part of the safety meeting.

Voluntary Industry Standards

For instructors who wish to refer to the Voluntary Industry Standards (VIS) for chemical process industries, Chapters 2–8 in the instructor's edition begin with a list of the VIS that pertain to the chapter's safety topic. The complete list of standards is published in *Foundations for Excellence in the Chemical Process Industries,* ISBN: 8412-3492-2, product code CT001. For information on ordering this book, contact the American Chemical Society, 800/227-5558, *www.acs.org/education/curriculum/*.

Please note that the VIS were drafted for the national context by ACS. The development of local and/or state standards is encouraged. Using the VIS as the starting point, local and state standards should be developed through collaboration with the affected employers. Many community colleges have the resources to guide such a collaboration without imposing traditional academic perspectives on the product.

Additional Resources

The ideas and information in this book provide a starting point for making safety education an integral part of your program. Many additional resources are available to further illustrate the topics presented in this book and reinforce the application of safety knowledge and skills in the workplace.

Many excellent safety videotapes are available for purchase from the American Chemical Society, 800/227-5558, *www.acs.org*. Also, the Laboratory Safety Institute, 508/647-1900, *www.labsafety.org*, has an audiovisual lending library that includes many of the ACS videos. These materials are available free of charge to members of the Laboratory Safety Institute and for a rental fee to nonmembers.

Representatives from local industry, chapters of the ACS Division of Chemical Technicians, and your institution's safety and health office can provide valuable expertise. Many would be happy to do presentations and/or provide examples of safety practices in their workplaces.

Keeping Up-to-Date

This book reflects current views of prudent laboratory practice and laws governing laboratory safety as of the publication date. Since the safety field encompasses so many topics and regulations that change frequently, staying up-to-date in all of them is very difficult. We recommend that you check periodically with your institution's safety and health office for important updates on safety laws or practices.

Chapter 1

The Culture of
Laboratory Safety

If good habits are inculcated from the beginning, participation in the culture of safety will be natural and painless.

Prudent Practices in the Laboratory, page 2

Chapter 1 Objectives

Students will learn

- the importance of a "safety culture" and ideas for establishing one,

- the responsibility professional laboratory workers have to maintain their own safety knowledge, and

- how to develop a safety library and locate Web resources of safety information.

Overview of the Chapter

Chapter 1: The Culture of Laboratory Safety is intended to provide students with a foundation for understanding the meaning of safety culture and its application to the academic and workplace settings.

To help model a safety culture for students, you may want to consider implementing one or more of the following ideas into your program: develop a safety program modeled after industry, including a safety committee, safety coordinator, safety meetings, intensive safety training for students from the first day of classes, and student orientation; integrate safety awareness into the program's promotional literature and advertising; add safety evaluation to student assignments, assessments, and performance evaluations; develop a safety reference library for student use; and integrate safety questions and assignments into all parts of chemistry classes. *Prudent Practices in the Laboratory* recommends that students actively consider risks, regulations, waste disposal, and costs associated with alternative approaches to problems under discussion.

References for Chapter 1

About NIOSH. www.cdc.gov/niosh/about.html (accessed March 15, 2000).

Geller, E.S. *The Psychology of Safety;* CRC: New York, 1998.

Geller, E.S. "Safety Coaching," *Professional Safety,* American Society of Safety Engineers, July 1995, p 16–22.

Hazard Communication Standard. *Code of Federal Regulations,* 29 CFR 1910.1200, 1999. (This is available free online at www.osha.gov.)

Hofstader, R.; Chapman, K. *Foundations for Excellence in the Chemical Process Industries: Voluntary Industry Standards for Chemical Process Industries Technical Workers;* American Chemical Society: Washington, DC, 1997.

Improving Safety in the Chemistry Laboratory: A Practical Guide, 2nd ed; Young, J.A., Ed.; John Wiley & Sons: New York, 1991.

Occupational Exposure to Hazardous Chemical in Laboratories Standard. *Code of Federal Regulations,* 29 CFR 1910.1450, 1999. (This is available free online at www.osha.gov.)

Chapter 1: The Culture of Laboratory Safety

If good habits are inculcated from the beginning, participation in the culture of safety will be natural and painless; if these lessons are neglected until . . . the first industrial job, reeducation can be a difficult, expensive, and perhaps even a dangerous initiation.

Prudent Practices in the Laboratory, page 2

The objective of this chapter is to provide you with a foundation for understanding the meaning of safety culture and its application to the academic and workplace settings.

In industrial and governmental laboratories, "safe" business is becoming synonymous with "good" business, and employers are increasingly active in developing and maintaining a culture of safety "where the safety culture is thoroughly understood, respected, and enforced from the highest level of management down." *(Prudent Practices) Prudent Practices* defines safety culture as "encompassing a group of people who voluntarily and willingly think about potential hazards and seek out and use resources that help ensure the maximum safe use of materials and procedures."

The Psychology of Safety uses the term "total safety culture," defining it as a workplace that incorporates the following mission and values:

- Everyone feels responsible for safety and does something about it on a daily basis.

- People go beyond the call of duty to identify unsafe conditions and at-risk behaviors, and they intervene to correct them.

- Safe work practices are supported with rewarding intermittent feedback from both peers and managers.

- People "actively care" continuously for the safety of themselves and others.

- Safety is not considered a priority that can be conveniently shifted depending on the demands of the situation; rather, safety is considered an immovable value.

A culture of safety was not always the norm in industry. According to *Prudent Practices,* "Until recently, the chemical hazards in many laboratories were not accepted and taken into account by those working in them, and, accordingly, the necessity of putting 'safety first' was not fully appreciated." *Prudent Practices* discusses several factors that are changing the culture of safety:

- Advances in technology have begun to change the safety requirements for chemical laboratories. These advances include miniaturizing chemical operations and simulating laboratory experiments by computer as an adjunct to laboratory work.

- The widely accepted goal of pollution prevention has also affected the culture of safety in laboratory work. The terms "waste reduction" and "waste minimization" are often used to discuss pollution prevention goals.

- Changes in the legal and regulatory climate have made collection and disposal of laboratory waste a major budget item in the operation of chemical laboratories. Additionally, protecting employees from toxic materials is both a moral obligation and an economic necessity.

The changes in safety culture have affected both academic and industrial laboratories, but industry standards and practices for health, safety, and environmental issues are typically much more stringent than those in academic laboratories. Industry human resource personnel report that new employees coming from academic laboratories often lack awareness of the detailed and rigorous safety procedures required by industry. In fact, those outside of industry may find it hard to believe what those at DuPont and other companies know: the accident rate at colleges and universities is 1,000 times as great as the accident rate at DuPont. In most companies today, working safely is a condition of employment. Employers do not want to have to break bad habits regarding safety that may have been allowed in academic training.

Part of developing and supporting a safety culture is having a clear written description of the safe attitudes and practices expected of all students and/or employees. For example, most laboratories have a chemical hygiene plan (CHP) that outlines the policies and procedures for working safely with hazardous materials. (The CHP is required for most laboratories under federal law. This topic is discussed in Chapter 8: Safety Planning and Standard Operating Procedures.)

Today, chemists, chemical technicians, chemical engineers, and others working in the chemical process and allied industries are expected, from their first day of employment, to understand and practice safe laboratory behaviors. Furthermore, all technical workers need to be able to apply their entry-level knowledge and skills to a broad variety of different laboratory operations.

Creating a Safety Culture

Major changes in both personal and corporate mind-sets are needed to enhance and universally effect the upgrading of safety that is being called for. According to *The Psychology of Safety,* the following shifts are needed to support a total safety culture:

1. *Shift from government-regulated safety programs to corporation-initiated programs:* Programs and documentation used to advertise and facilitate them should embrace the proactive safety policies as an integral part of the corporate infrastructure rather than a government-driven effort.

2. *Shift from failure-dominated safety goals to achievement-oriented goals:* Safety goals should not be stated in negative reinforcement terms (such as measuring safety with records of losses and injuries), but rather in positive achievement terms (such as tracking increases in safe behavior).

3. *Shift from outcome-focused incentives to behavior-focused incentives:* Incentives should not be based on outcomes (such as number of days without injuries), but rather on the process needed to reach the outcome (such as increasing safe behaviors and decreasing unsafe behaviors). Incentives based on fewest injuries can often reduce the reported numbers without improving safety, as employees (or students) may feel pressure to cover up an accident or injury in order to "win."

4. *Shift from top-down involvement to bottom-up efforts:* Large-scale and long-term behavior change must be supported from the top but driven from the bottom up, as employees (and students) themselves apply safety techniques throughout their workplace.

5. *Shift from rugged individualism to teamwork:* An individualistic, "I win/you lose" perspective must be replaced with teamwork founded on interpersonal trust and win/win contingencies. This is a challenge, as grades in school, the legal system, many sports, promotions, and many other social structures orient us to a win/lose mind-set.

6. *Shift from a piecemeal approach to a systemic approach:* A total safety culture requires a systemic approach with balanced attention to three domains:

   • environmental factors (including equipment, tools, physical layout, and standard operating procedures);

   • person factors (including people's knowledge, skills, attitudes, beliefs, and personalities); and

   • behavior factors (including safe and at-risk work practices, communicating, and demonstrating active care for others' safety).

7. *Shift from fault-finding to fact-finding:* Blaming individuals or groups for an incident is not consistent with a systemic approach. Instead, an accident, injury, or near-miss is an opportunity to investigate facts from all aspects of the system (environment, person, and behavior) that could have contributed to the incident.

8. *Shift from reactivity to proactivity:* Investigating the events leading up to an accident points out where proactive thinking and acting could have prevented the incident, but time to be proactive is difficult to find. (See next item.)

9. *Shift from quick fix to continuous improvement:* No quick fix can establish a safety culture or improve safety on a large scale. With an emphasis on continuous improvement, eventually "proactive" can replace "reactive."

10. *Shift from a changeable priority to an immovable value:* "Safety is a Priority" is one of the most common safety slogans. Geller says that safety should not be a priority, but rather a value, because priorities can be arranged for our convenience while values are much more immovable. He illustrates this point by describing the morning routines of various people. Some eat breakfast, others exercise. Some just grab coffee and go out the door. Each person has different priorities. Yet, one common activity is not a priority but a basic value—if each of these people had only 15 minutes to prepare for work, their priorities would be rearranged, but every one of them would do one thing that represents a value taught from infancy: they would get dressed. Geller emphasizes that safety should be a value that is just as strong, not a priority that can take a "back seat" when other priorities, such as deadlines, take precedence.

Safety Coaching in Support of a Safety Culture

An institution's dedication to safety does not guarantee safe behavior in its employees (or students). Complacency, apathy, or even active resistance to safety rules and procedures must be combated to achieve a successful safety culture in the laboratory. Many safety consultants recommend drawing on the principles of psychology to improve attitudes and behaviors about workplace safety. The safety coaching process described in this section is a behavior-based approach that has been successful in industry settings and that translates readily into classroom settings.

The Psychology of Safety recommends organizing a team of peers called "safety coaches" as an important step to creating and maintaining a total safety culture within the company. The safety coaches objectively observe and analyze the working practices of co-workers, recognizing and supporting safe behavior while offering constructive feedback about unsafe behavior. In some companies, workers elect their own safety coaches or take turns serving this function. "Safety coaching must not be viewed as a way to enforce rules or 'catch' an employee making errors; rather coaching should be perceived as a process designed to help the employee develop safe work habits via supportive and corrective feedback." Geller states that "people are often unwilling to coach or be coached for safety because they view safety from an individualistic perspective." People need to consider safety coaching a shared responsibility. According to Geller, the five letters of the word COACH can be used to remember the basic ingredients of the most effective coaching:

Care: Safety coaches must truly care about the health and safety of their co-workers (or fellow students) and show that caring though words, body language, and actions.

Observe: Safety coaches must observe the behavior of others systematically and objectively. The observer should never hide or spy and should always ask permission. The observer also needs to use an agreed-upon list of critical behaviors. (See "Example Critical Behavior Checklist" on the next page.)

Analyze: Safety coaches must understand that risky behaviors occur for a variety of reasons, including work demands, example-setting by peers, inconsistent messages from management, comfort, and convenience. This understanding is essential to ensure that safety coaching serves as a "fact-finding" and not a "fault-finding" process. An objective analysis can help find the reasons behind at-risk behaviors and help develop interventions to decrease them.

Communicate: A good coach must be an active listener and a persuasive speaker. Most of us need practice to develop and effectively use these skills. Coaches must be able to separate behavioral feedback from personal factors such as attitudes and feelings so that the person being coached does not feel defensive.

Help: The purpose of being a safety coach is to help co-workers or fellow students work more safely.

A critical behavior checklist (CBC) is an essential part of safety coaching. The first step is to decide what items to include on the CBC. The list could quickly grow overwhelming, so a CBC should include only behaviors that 1) have caused or could cause a large number of injuries or near-misses or 2) have caused or could cause a serious injury or fatality. Another important criterion in selecting critical behaviors is that they must be objective and observable. Some organizations encourage employees to develop these checklists in order to build feelings of ownership and commitment. The format and organization of CBCs varies with each organization, but the coach should have a place to checkmark the safe and at-risk behaviors observed as well as to write down comments. A CBC might include the behaviors described in the following table:

Example Critical Behavior Checklist		
Operating Procedure	Safe Observation	At-Risk Observation
Using goggles and other personal protective equipment (PPE). The PPE is appropriate for the chemicals being handled: for example, appropriate laboratory gloves are used.		
Dressed appropriately: long hair is contained. No loose-fitting clothes. No open-toed shoes.		
Checks to see if laboratory bench surface is uncluttered and free from spills, and if not, takes appropriate action.		
Checks labels on chemical containers before using.		
Chemicals are properly disposed of after use.		
Adapted with permission from Geller, E.S., *The Psychology of Safety;* CRC: New York, 1998.		

Other issues to consider when developing CBCs include the following:

- A CBC is a continuous improvement process; further development and refinement can enhance coaching effectiveness.

- The list of critical behaviors should be kept short at first. An exhaustive checklist may appear overwhelming and inhibit progress.

- A work group developing a CBC must meet periodically to select and refine critical behaviors to be observed.

With workers' input, observation sessions should be scheduled to fit the work setting and process. Protocols for scheduling effective coaching observations vary widely. Geller provides two examples. One company has each employee schedule "two coaching sessions with any two employees each month. Thus on days and at times selected by the observee, two observers use a CBC to conduct a 30-minute session." This procedure is markedly different than the planned 60-second review implemented at another company. "In this process, all employees attempt to complete a one-minute daily observation of another employee's safe/unsafe work practices."

Using a CBC, a safety coach can calculate a "percent safe behavior" for each observation session. This gives the coach, the person being observed, and the group as a whole a tangible and positive measure of improvement. After each observation session, the results are usually calculated by inserting the number of safe observations and at-risk observations into the following equation to yield "percent safe behavior":

$$\text{Percent Safe Behavior} = \frac{\text{Total Safe Observations}}{\text{Total Safe Observations} + \text{At-Risk Observations}} \times 100$$

The key to the success of this approach is ensuring that participants understand that the process is more important than the numbers. The ultimate goal of the coaching process is to change behavior. By understanding the context of the at-risk behavior, the safety coach can demonstrate that the CBC process is not about finding fault with the observee but about finding solutions for increasing safety. Observers and coaches must understand that anyone can unconsciously act unsafely and that performance can only improve with behavior-specific feedback. To this end, safety coaches should aspire to a specific set of skills that will strengthen their ability to provide effective feedback. These include having a genuine caring attitude for the safety of co-workers, a strong sense of humor, sensitivity to the self-esteem of co-workers, and the ability to listen and give praise to others.

Some other tips for coaching include the following:

- Give feedback one-on-one and privately.

- Give feedback as soon as possible after the observation process.

- Identify the safe behavior(s) observed.

- Identify any unsafe behavior observed and indicate concern for the person's welfare.

- Be sincere and genuine.

- Thank the person for his or her commitment to continuous improvement.

Use of Safety Coaching and CBCs in the Classroom and Laboratory

Many of *The Psychology of Safety's* procedures are adaptable to the classroom as exercises or activities that can be conducted during or after covering a topic in this book. You may want to discuss with your instructors a plan for incorporating the development and use of CBCs into your coursework. A suggested process follows:

1. Work with your classmates to develop a list of safe and at-risk behaviors in the laboratory. Ideas may come from chapters in this book, laboratory standard operating procedures, accident records, individual anecdotes, and other resources.

2. Work together to pare down the lists into a small set of the most critical behaviors. This is a CBC. You may also want to develop several small CBCs that might be used in different situations.

3. Test the observability of each behavior in the CBC by taking turns modeling safe and unsafe behaviors. Test to see whether independent observers can agree on their observations.

4. Revise the CBC if necessary.

5. Determine whether a given behavior is best measured by frequency or duration.

6. Agree on the frequency of observation (whether observations should be scheduled for specific periods of time or occur at any time).

7. Appropriate feedback for safety coaching should emphasize humor, empathy, and listening ability. Take turns modeling or role playing appropriate and inappropriate feedback, and critique each other's performances. Remember that unsafe behaviors can occur for many reasons and that feedback about unsafe behavior is not an insult to the person being observed.

Developing Your Safety Knowledge and Skills

In a safety culture, you are ultimately responsible for your own safety and for the safety of those around you, and you make every effort to fulfill that responsibility. Industry expects new hires to know how to work safely in a laboratory from day one on the job. Excelling in this area can be the basis for promotion and financial reward. To prepare to excel in industry, start now to proactively identify and follow your institution's standard operating procedures (SOPs); all local, state, and federal regulations; and generally accepted standards on safe handling and storage of all chemicals and equipment you may use in labs. This includes determining and using the appropriate personal protective equipment (e.g., goggles, gloves, apron). Many print and online resources are available to assist you in obtaining the information you need. The background information in this book and the glossary at the end will help you begin your safety education.

For more in-depth information, the texts listed here would serve as the foundation for a good safety library:

Armour, M.A. *Hazardous Laboratory Chemicals Disposal Guide,* 2nd ed.; Lewis: Boca Raton, FL, 1996.

CRC Handbook of Laboratory Safety, 4th ed.; Steere, N.V., Ed.; CRC: Boca Raton, FL, 1995.

Committee on Chemical Safety. *Chemical Safety Manual for Small Businesses, Guide for Managers, Administrators and Employees;* American Chemical Society: Washington, DC, 1989.

Compressed Gases: Compressed Hazards [video]; American Chemical Society: Washington, DC, 1995.

Compressed Gases: Safe Handling Procedures [video]; American Chemical Society: Washington, DC, 1995.

Coyne, G.S. *The Laboratory Companion: A Practical Guide to Materials, Equipment, and Technique;* John Wiley & Sons: New York, 1997.

Fire Protection Guide on Hazardous Materials; National Fire Protection Association: Quincy, MA, 1999.

Geller, E.S. *The Psychology of Safety;* CRC: New York, 1998.

Gorman, C. *Working Safely with Chemicals in the Laboratory,* 2nd ed.; Genium: Schenectady, NY, 1997.

Hall, S.K. *Chemical Safety in the Laboratory;* Lewis: Boca Raton, FL, 1994.

Hazard Communication Standard. *Code of Federal Regulations,* 29 CFR 1910.1200, 1999. (This is available free online at www.osha.gov.)

Improving Safety in the Chemical Laboratory: A Practical Guide, 2nd ed.; Young, J.A., Ed.; John Wiley & Sons: New York, 1991.

Kingsley, W.K., Segal, E.B., Phifer, R. *Living with the Laboratory Standard, A Guide for Chemical Hygiene Officers;* Committee on Chemical Safety, American Chemical Society: Washington, DC, 1998.

Laboratory Safety Guidelines; Laboratory Safety Institute. www.labsafety.org/40steps.htm (accessed March 3, 2000).

Learning By Accident; Mojtabai, F., Kaufman, J., Eds.; Laboratory Safety Institute: Natick, MA, 1997.

Learning By Accident, Vol. 2; Mojtabai, F., Kaufman, J., Eds.; Laboratory Safety Institute: Natick, MA, 2000.

Mercier, P. *Laboratory Safety Pocket Handbook;* Genium: Schenectady, NY, 1996.

NIOSH Pocket Guide to Chemical Hazards; National Institute of Occupational Safety and Health: Washington, DC. (The NIOSH guide is available in print, on CD-ROM, and on diskette from National Technical Information Service, 800/553-6847, orders@ntis.fedworld.gov, order #PB 97-177-604.)

Occupational Exposure to Hazardous Chemicals in Laboratories Standard. *Code of Federal Regulations,* 29 CFR 1910.1450, 1999. (This is available free online at www.osha.gov.)

Pocket Guide to MSDSs and Labels; Business and Legal Reports: Madison, CT, 1990.

Prudent Practices in the Laboratory; National Academy: Washington, DC, 1995.

Safe Handling of Compressed Gases in Containers, 8th ed.; Compressed Gas Association: Arlington, VA, 1991.

Safe Handling of Compressed Gases in the Laboratory and Plant; Matheson Gas Products: Chicago, IL, 1985. (This is available free from Matheson Gas 800/828-4313 x309.)

Safety in Academic Chemistry Laboratories; American Chemical Society: Washington, DC, 1998.

Shugar, G.J.; Shugar, R.A.; Bauman, L.; Bauman, R.S. *Chemical Technicians' Ready Reference Handbook,* 4th ed.; McGraw-Hill: New York, 1996.

Task Force on Laboratory Waste Management. *Laboratory Waste Management: A Guidebook;* American Chemical Society: Washington, DC, 1994.

Wood, C. *Safety in School Science Labs;* Kaufman and Associates: Natick, MA, 1991.

Young, J.A.; Kingsley, W.K.; Walsh, G.H. *Developing a Chemical Hygiene Plan;* American Chemical Society: Washington, DC, 1990.

A recommended list of websites includes the following:

- *www.osha.gov/* The Occupational Safety and Health Administration's (OSHA) website contains information about the agency, news releases, fact sheets, and the complete list of OSHA Regulations (Standards–29 CFR).

- *www.cdc.gov/niosh/homepage.html/* NIOSH's website contains numerous online documents including the Pocket Guide to Chemical Hazards and the Certified Equipment List.

- *www.acgih.org/* The American Conference of Governmental Industrial Hygienists, Inc. (ACGIH) offers a list of available publications and a comprehensive set of links to allied organizations.

- *hazard.com/* This website provides over 180,000 online searchable material safety data sheets, plus other safety links and information.

- *www.osh.net/* This site features about 60 informational categories that cover nearly all safety and health disciplines. The site has a health-and-safety message board where users can post inquiries and messages.

What specific safety knowledge, skills, and attitudes do you need in order to be prepared for the workplace? The Voluntary Industry Standards, or VIS, are a set of competencies for worker knowledge and skills developed in response to concerns about worker skills in general and are intended to ensure a skilled work force. The US Departments of Education and Labor have worked together with industry to develop the VIS in a variety of industries which can then be communicated to faculty and students in these fields. The American Chemical Society developed the VIS for chemical process industries and published them in a book titled *Foundations for Excellence in the Chemical Process Industries* (ISBN: 8412-3492-2). Many of the VIS competencies are directly pertinent to safety. If you would like to see the VIS for chemical workers, check your school library or contact the American Chemical Society, 800/227-5558, *www.acs.org/education/curriculum/.*

Chapter 2
Precautionary Labels

A friend was working in the laboratory using several organic solutions. They were not labeled. They got mixed up and an explosion resulted. My friend was burned and severely cut by the glass particles. As a result he lost the sight in his right eye.

Learning by Accident, page 73

Chapter 2 Objectives

Pertinent Voluntary Industry Standards

- *VIS L2 Tasks*—Labeling all chemicals, materials, tools, and equipment with appropriate safety, health, and environmental details.

- *VIS L2 Tasks*—Ensuring that warning labels are displayed appropriately.

- *VIS (L2.01):07*—Describe the Department of Transportation (DOT) regulations of labeling and shipping hazardous wastes; including the possibility of personal liability.

- *VIS (L2.02):02*—State the responsibilities and rights of the technician under the Hazard Communication Standard of the Occupational Safety and Health Administration (OSHA).

- *VIS (L2.02):04*—Identify the conventions and symbols used for labeling chemical materials; include Hazardous Material Identification System (HMIS) and National Fire Protection Association (NFPA) guidelines.

- *VIS (L2.02):07*—Demonstrate the ability to read, interpret, and prepare labels for a variety of chemical materials.

- *VIS L3 Standards*—Demonstrate ability to correctly label and store chemicals of all types.

- *VIS (L3.02):07*—Apply various coding systems used for describing the properties of compounds that may be important in hazardous conditions (i.e., Diamond).

Goals of the Chapter

Students will learn

- what information is provided on consumer and industrial chemical labels,

- the hazard identification systems used by the HMIS and NFPA,

- the use of safety pictograms on chemical labels, and

- labeling procedures for chemical containers in the laboratory and shipping containers.

Why Your Students Need to Understand Precautionary Labels

Under the OSHA Hazard Communication Standard (29 CFR 1910.1200), workers have a right to know about the potential hazards of the chemicals they may encounter in their workplace. Chemical labels are the primary form of hazard communication, immediately conveying to the user any hazards and potential risks associated with a chemical. OSHA requires that chemicals entering the workplace be appropriately labeled and that the labels remain intact.

In the workplace, your students will depend on precautionary labels from chemical suppliers for critical information, and thus they need to be taught how to routinely read and follow the precautions on such labels. Although OSHA does not require a specific design for precautionary labels, OSHA (with reference to the American National Standards Institute, ANSI) does require that certain information appear. The design of the label often incorporates various codes, symbols, and abbreviations. Your students will benefit from knowing what information to expect on precautionary labels and being familiar with the codes and symbols they will routinely see.

Preparing chemicals for shipment is another important labeling issue. In the workplace, your students may have to package and label chemicals for return to a supplier or to send to another laboratory for testing. These packages must be prepared according to very specific Department of Transportation (DOT) requirements.

Although not currently regulated by OSHA, good laboratory practice includes labeling storage, dispensing, and temporary containers in which chemicals are stored or held while in use. In the workplace, your students will be responsible for consistently using appropriate labels on these kinds of containers.

Overview of the Chapter

Chapter 2: Precautionary Labels contains four sections. General background information and complete instructions for the exercises are provided. The background pages are intended only to provide basic background information on labels for your students. We recommend that you extend this background information by having students find and read the pertinent sections of the books in the safety reference library (described in Chapter 1). You may wish to have students briefly report on the information they find during the monthly safety meetings.

Section 2A: Getting to Know Labels

This section introduces students to the importance of chemical labels and the regulations governing their use in the laboratory and workplace. To initiate interest in the topic, students read and discuss Laboratory Incident Reports concerning accidents that resulted from the use of unlabeled containers or from labels being ignored or misread.

Section 2B: Chemical Labels in the Laboratory and at Home

Students critically review industrial and consumer product labels for four chemicals: 3% H_2O_2, household bleach, rubbing alcohol, and toluene. The objective is for students to understand the role that labels play in keeping users informed, how industrial and consumer product labels differ, and how being informed about a chemical can help prevent accidents.

Section 2C: Looking at Label Codes and Symbols

In this section students become familiar with and evaluate the advantages and disadvantages of common codes and symbols, including the National Fire Protection Association (NFPA) diamond, the Hazardous Material Information System (HMIS) bars, and common pictograms.

Section 2D: Labeling Containers

The objective of this section is for students to understand the requirements for proper labeling of chemicals being stored in something other than their original container, being placed in a temporary container during an experiment, or being packed for shipment. Students use labels from commercially packaged products as references to practice making new labels for containers used in house for storing and dispensing chemicals.

References for Chapter 2

EM Science Catalogue, EM Science: Cincinnati, OH, 1998.

Fisher Catalogue, Fisher Scientific: Pittsburgh, PA, 1995–1996.

Hazard Communication Standard. *Code of Federal Regulations*, 29 CFR 1910.1200, 1999. (This is available free online at www.osha.gov.)

Hazardous Industrial Chemicals—Precautionary Labeling. *American National Standards Institute Standards,* ANSI Z129.1-1994, American National Standards Institute: Washington, DC.

Hazardous Material. *National Fire Protection Association Standards*, Title 1, Part 5, Section 704.

Hazardous Materials Procedure. *Code of Federal Regulations*, 49 CFR 100–185, 1999. (This is available free online at www.gpo.gov.)

Hofstader, R.; Chapman, K. *Foundations for Excellence in the Chemical Process Industries: Voluntary Industry Standards for Chemical Process Industries Technical Workers;* American Chemical Society: Washington, DC, 1997.

Improving Safety in the Chemical Laboratory: A Practical Guide, 2nd ed.; Young, J.A., Ed.; John Wiley & Sons: New York, 1991.

Labeling Requirements; Prominence, Placement, and Conspicuousness Standard. *Code of Federal Regulations*, 16 CFR 1500.121, 1999. (This is available free online at www.gpo.gov.)

Langerman, N. *Precautionary Labels for Chemical Containers;* Lewis: Boca Raton, FL, 1994.

Learning By Accident; Mojtabai, F.; Kaufman, J., Eds.; Laboratory Safety Institute: Natick, MA, 1997.

Mercier, P. *Laboratory Safety Pocket Guide;* Genium: Schenectady, NY, 1996

Paint and Coatings Standard. *Hazard Materials Information System;* National Paint and Coatings Association: Washington, DC, 1995.

Pocket Guide to MSDSs and Labels; Business and Legal Reports: Madison, CT, 1990.

Prudent Practices in the Laboratory; National Academy: Washington, DC, 1995.

Purpose and Use of Hazardous Materials Table. *Code of Federal Regulations,* 49 CFR 171.101,1999. (This is available free online at www.gpo.gov.)

Safety in Academic Chemistry Laboratories; American Chemical Society: Washington, DC, 1998.

Samoyault, T. *Give Me a Sign: What Pictograms Tell Us Without Words;* Viking: New York, 1997.

Section 2A: Getting to Know Labels
Instructor Notes

This section introduces students to the importance of chemical labels and the regulations governing their use in the laboratory and workplace. To initiate interest in the topic, students read and discuss Laboratory Incident Reports concerning accidents that resulted from the use of unlabeled containers or from labels being ignored or misread.

Student Pages

- Section 2A: Getting to Know Labels—Laboratory Incident Reports
- Section 2A: Getting to Know Labels—Background

Possible Playout of the Section

Students read the Laboratory Incident Reports and Background and discuss the following questions:

1. What are the similarities and differences between the Laboratory Incident Reports?

2. What conclusions can you draw about safety from these incidents?

3. How might each of these incidents have been prevented?

Section 2A: Getting to Know Labels
Laboratory Incident Reports

Incident Descriptions: Unlabeled Chemicals

A friend was working in the laboratory using several organic solutions. They were not labeled. They got mixed up and an explosion resulted. My friend was burned and severely cut by the glass particles. As a result he lost the sight in his right eye.

Learning by Accident, page 73

I was getting an experiment ready involving bleach. I took a small amount out and put it along with a medicine dropper in an unlabeled beaker. A student came in and waved the dropper in the air, thinking it was water. I got a few drops in my eye and ended up with corneal scratches and an eye patch.

Learning by Accident, page 73

Incident Descriptions: Not Reading Labels

After having eggs thrown at her house on mischief night, a [woman from] Portland, Maine . . . went to clean off the remains the next day. While outside, she mixed a solution of household ammonia and Clorox® Bleach. She was killed. Someone saw her pass out and called an ambulance. One attendant was overcome and later died.

Learning by Accident, page 11

A technician was recrystallizing an organic compound from boiling acetic acid. Needing more solvent, she went to the end of the bench for another bottle of acetic acid. Suddenly there was a loud "whoosh" and an eruption of boiling acid hit the ceiling. I was very puzzled to explain the event until I remembered noticing a cloud of brownish fumes immediately after the eruption. An examination revealed that the technician had taken a bottle of concentrated nitric acid instead of the acetic acid. Fortunately, there were no injuries.

Learning by Accident, page 31

A class was doing an experiment on enzymes. Hydrogen peroxide was added to some liver and a gas was supposed to be released. Students supposedly added 3% hydrogen peroxide to the food in a test tube and were instructed to shake the test tube. They stoppered the tubes with their thumbs. In no time students complained of burning sensations on their skin. About 10 students suffered burns before the activity was stopped. The teacher had the students wash their hands profusely. They went to the school nurse and the local hospital. The problem seemed to be that 30% hydrogen peroxide had been dispensed instead of 3%. The hydrogen peroxide had been taken from a large stock bottle stored in the laboratory refrigerator. The label had not been carefully read.

Learning by Accident, page 25

Section 2A: Getting to Know Labels

Background

Note: This section provides general recommendations for and information about chemical labels. However, you should always follow the procedures established by your organization's chemical hygiene plan.

Chemical labels are designed with user safety in mind, whether for workers in an industrial laboratory or household consumers. These labels are the primary form of hazard communication, conveying to the user any hazards and potential risks associated with a chemical. *Precautionary Labels for Chemical Containers* states, "The warnings should be based on the inherent properties of the chemical or mixture as well as potential hazardous exposures resulting from reasonably foreseeable occupational use, handling, storage, and misuse. The language used on a chemical label should contain instructions that, if followed, are sufficient to assure that no injury or illness will occur as a result of handling the chemical."

The information that can feasibly be included on a label is limited. While size is one obvious factor, readability is another important consideration. Labels must grab the reader's attention and provide enough information for readers to use the product safely. Labels should not be too chatty or provide so much information that they are superficially skimmed or ignored.

Consumer labels that appear on products such as bleach, drain cleaner, and paint thinner have several basic elements in common: the identity of the product, directions for use, precautionary statements (including signal words such as DANGER), how to avoid exposure, and what to do in case of exposure. This information, while adequate for home use, may not be complete enough for the industrial workplace. Also, different regulations govern the content of consumer and industrial labels. *You will examine the content of consumer and industrial labels in Section 2B: Chemical Labels in the Laboratory and at Home.*

Codes and Symbols Used for Labeling Chemicals

Various codes and symbols are used as components of chemical labels. These include a variety of shapes, pictures, colors, numbers, and abbreviations that you must be able to recognize. A frequently seen set of codes and symbols is the National Fire Protection Association (NFPA) diamond. The NFPA diamond uses a combination of colors (blue for health, red for flammability, yellow for reactivity, and white for other), a rating scale of 0 to 4, and other symbols. Another set of codes and symbols is the Hazardous Material Information System (HMIS). The HMIS uses bars of color (blue for health, red for flammability, yellow for reactivity, and white for protective equipment), a rating scale, and pictograms for appropriate protective equipment, such as gloves and goggles. These codes and symbols provide only part of the required information, so they cannot be used as a complete label. Pictograms are often used to provide quick recognition of hazard information. According to *Precautionary Labels for Chemical Containers,* "pictograms convey more information than any other label element." *You will become familiar with and practice using NFPA and HMIS-type symbols in Section 2C: Looking at Label Codes and Symbols.*

Regulations Governing Labels

The Occupational Safety and Health Administration (OSHA), the U.S. Environmental Protection Agency (EPA), the Consumer Product Safety Commission (CPSC), and the U.S. Department of Transportation (DOT) all have standards for communication of potential hazards through labels. The OSHA standard also cross-references the American National Standards Institute (ANSI) Labeling Standard, which provides additional recommendations.

OSHA does not regulate the labeling of very small containers, such as the beakers and test tubes found in laboratories, but it is a good practice to label all containers and glassware containing chemicals. Such labeling organizes the laboratory, communicates activities to others working in the laboratory, and helps you keep track of laboratory chemicals that you are using. Since many liquids are clear and colorless, without labels they are virtually impossible to identify. In industry, the mechanism of laboratory labeling is left to be developed by the employer as part of the Chemical Hygiene Plan. *Prudent Practices in the Laboratory,* a laboratory safety reference developed by the National Research Council, provides suggestions for the labeling of laboratory containers. *You will become familiar with these recommendations and will practice labeling small laboratory containers in Section 2D: Labeling Chemicals in the Laboratory.*

All hazardous materials being transported away from facilities must comply with labeling regulations to protect the public, those transporting the materials, and personnel receiving the materials. The DOT requires the use of shipping labels on cartons and packages containing chemicals identified in the Hazardous Materials Table section of 49 CFR 172.101. Facilities shipping out these chemicals are required to identify, describe, and communicate the hazards on the shipping packages according to hazard classes defined in the Hazardous Materials Table. The labels required by the DOT do not provide full information about the hazards. They are meant to identify only immediate hazards that pose a risk during transportation.

Using Chemical Labels

Always read and understand the label on any chemical, whether in the laboratory or at home. Before you use the chemical, know what it is, what hazards exist, how to avoid those hazards, how to recognize when you have been exposed, and what to do if you have been exposed. *Precautionary Labels for Chemical Containers* states

> "Because much chemical usage is repetitive and because people become careless doing repetitive tasks, protective habits must be developed to compensate for human fallibility. The primary habit related to product identification is to identify the product at least three times: once before removing the container from storage, once before opening the container, and once before introducing the product into the process."

Never use a chemical from a container that is not labeled even if you feel reasonably sure that you know the chemical's identity. Such a situation is an accident waiting to happen. Alert your co-workers and your supervisor of the missing label, and take the container out of service until someone tests, identifies, and properly labels the container.

Labels and Material Safety Data Sheets

Material safety data sheets (MSDSs) are comprehensive descriptions of chemical properties and hazards required by OSHA for all hazardous chemicals. The chemical label and the MSDS are complementary aspects of the hazard communication process, and both must be used to handle chemicals safely. Many chemical labels reference the MSDS for the specific chemical in question. *You will learn about MSDSs in Chapter 3: Material Safety Data Sheets.*

Section 2B: Chemical Labels in the Laboratory and at Home

Instructor Notes

Students critically review industrial and consumer product labels for four chemicals: 3% H_2O_2, household bleach, rubbing alcohol, and toluene. The objective is for students to understand the role that labels play in keeping users informed, how industrial and consumer product labels differ, and how being informed about a chemical can help prevent accidents.

Student Pages

- Section 2B: Chemical Labels in the Laboratory and at Home—Background
- Section 2B: Chemical Labels in the Laboratory and at Home—Exercise

Possible Playout of the Section

Students read the Background and then compare the industrial and consumer product labels in the Exercise. They discuss differences between the two types of labels, including intended audiences, purposes, and language styles. Ensure that students understand the specific details provided in these labels by having them answer the following questions or by having a class discussion centered around these questions:

1. What precautions, special handling, and/or protective equipment are required for each?

2. What is the single most hazardous situation posed by each of these chemicals?

3. What should one do in case of contact, exposure, spill, or leak?

Section 2B: Chemical Labels in the Laboratory and at Home

Background

Note: This section provides general recommendations for and information about chemical labels. However, you should always follow the procedures established by your organization's chemical hygiene plan.

The function of laboratory and household chemical labels is to communicate information about chemical hazards and potential risks and actions to be taken in case of exposure. These types of labels have similarities and differences, as discussed below.

Understanding Signal Words

Signal words are used on both consumer and industrial labels. The signal words DANGER!, WARNING!, and CAUTION! warn users of immediate hazards associated with using a product. The use of these words is governed by detailed criteria established by the American National Standards Institute (ANSI). The general meaning of each signal word is described below. The selection of a signal word to be used on a label is based on the most serious overall immediate risk of using the product.

Signal Word	General Meaning
DANGER	If the product gets in you or on you, immediate harm will be caused.
WARNING	If the product gets in you or on you in sufficient quantity, you will suffer harm.
CAUTION	If this product gets in you or on you in large quantity over an extended period of time, you may be harmed.

Other words are used for delayed hazards, such as cancer or reproductive problems. Depending on the severity of risk, words for cancer hazards include CANCER HAZARD, SUSPECTED CANCER HAZARD, and POSSIBLE CANCER HAZARD. Words for reproductive hazards include BIRTH DEFECT HAZARD, POSSIBLE BIRTH DEFECT HAZARD, TERATOGEN, and POSSIBLE TERATOGEN. The term POISON, while not an ANSI signal word, is used on labels to indicate that the product will make you ill if it enters your body.

Consumer Product Labels

The content of consumer labels for household chemicals is governed by the Consumer Product Safety Commission (CPSC) and, for some specific chemicals, by the Environmental Protection Agency (EPA). According to CPSC regulations (16 CFR 1500.121), the label must prominently state the following information in conspicuous and legible type that contrasts in typography, layout, or color with other printed matter on the label:

- the name and place of business of the manufacturer, packer, distributor, or seller;

- the common or usual name, or the chemical name if there is no common or usual name, of the hazardous substance or of each component that contributes substantially to its hazard;

- the signal word DANGER on substances that are corrosive, extremely flammable, or highly toxic, as defined by 16 CFR 1500.3;

- the signal word WARNING or CAUTION on all other hazardous substances;

- an affirmative statement of the principal hazard or hazards, such as "Flammable," "Combustible," "Vapor Harmful," "Causes Burns," "Absorbed Through Skin," or similar wording descriptive of the hazard;

- precautionary measures describing the action to be followed or avoided;

- instruction, when necessary or appropriate, for first-aid treatment;

- the word POISON for any hazardous substance that is defined as "highly toxic" by CPSC in addition to the signal word DANGER and the skull-and-crossbones symbol;

- whether the chemical is a volatile organic compound (VOC) (required by EPA);

- instructions for handling (including personal protective equipment) and storage of packages that require special care in handling or storage; and

- (1) if the article is not intended for use by children, the statement "Keep out of the reach of children" or its practical equivalent, or (2) if the article is intended for use by children and is not a banned hazardous substance, adequate directions for protecting children from the hazard.

Industrial Chemical Product Labels

Industrial chemical suppliers are required to label their products according to the following guidelines, stated in OSHA regulation 29 CFR 1910.1200:

- state the identity of the hazardous chemical(s);

- provide appropriate hazard warning statements and/or words, pictures, and symbols that provide at least general information regarding the hazards of the chemicals, including a list of target organs potentially affected by the chemical;

- state the name and address of the manufacturer;

- provide legible symbols, pictures, and letters; and

- present information in English.

OSHA provides interpretive documents to clarify the meanings of these regulations, and these documents must be consulted to have a complete understanding of the requirements. The ANSI labeling standard (Z129.1-1994) provides additional useful information and should be consulted when developing labels. ANSI recommends the following items be included on industrial labels:

- identity of the product and hazardous constituents (if a mixture) or the identity of the pure chemical;

- signal word, such as DANGER, and a supplementary word, such as POISON;

- statement of hazard, such as "may be fatal if inhaled or swallowed";

- precautionary measures, such as "use only with adequate ventilation";

- instructions in case of contact or exposure, such as "in case of contact, immediately flush eyes or skin with plenty of water for 15 minutes";

- antidotes;

- notes to physicians;

- instructions in case of fire, spill, or leak;

- instructions for container handling and storage; and

- other useful information, such as Chemical Abstracts Service (CAS) Registry Number and reference to MSDS.

Because a given chemical can be referred to by more than one name (isopropyl alcohol and isopropanol, for example), using a simple, unique identifier for every chemical reduces the possibility of confusion. CAS provides such an identifier. In the mid-1960s, CAS developed the CAS Chemical Registry, a computer-based system that automatically identifies chemical structural diagrams and assigns to them a unique CAS Registry Number. This number, which has no chemical significance, is then used within a larger processing system to link the molecular structure with its Chemical Abstracts Index Name and other data. The CAS Registry Number, which may have up to nine digits, is divided into three parts separated by hyphens. The first part, starting from the left, has up to six digits. The second part has two digits, and the third and final part consists of a single check digit, which is used to verify by computer the validity of the entire number. *Prudent Practices in the Laboratory* recommends that laboratory staff add the CAS Registry Number as an unambiguous identifier to any commercial label that does not include it.

Formats for Industrial Chemical Labels

OSHA does not require a specific format for chemical labels in industry, so labels from different chemical suppliers may look quite different from each other. Generally, a company will use a consistent design for all of its labels to create a sense of continuity and enable users to find important information quickly and easily. (See Figure 2B-1.) Some of the information on these labels meets the regulatory requirements just discussed. Additional information about the product (such as grade, lot number, and lot analysis) is often provided.

❶ Name Block: Chemical name in large or bold letters. May include additional information such as chemical grade, structural or molecular formula, or formula weight.

❷ Specifications or Lot Analysis: Maximum limits of impurities or lot analysis.

❸ Storage: Suggested storage code (i.e., FLAMMABLE or REACTIVE); may be color-coded.

❹ Health and Safety: Health, Flammability, Contact, or Reactivity hazard warnings. May be a numerical code (0–4, from non- to extremely hazardous) or pictograms representing hazards or personal protective equipment.

❺ Written Precautions/First Aid: Clear, concisely written precautions and first aid instructions.

❻ NFPA Diamond Code: National Fire Protection Association symbol indicating the hazards a chemical poses during a fir

❼ UN or NA Number: United Nations or U.S. Department of Transportation coding system for hazardous materials.

❽ Target Organ: Warning information providing information regarding organs affected by exposure to the chemical.

❾ CAS Registry Number: Chemical Abstract Service Registry Number.

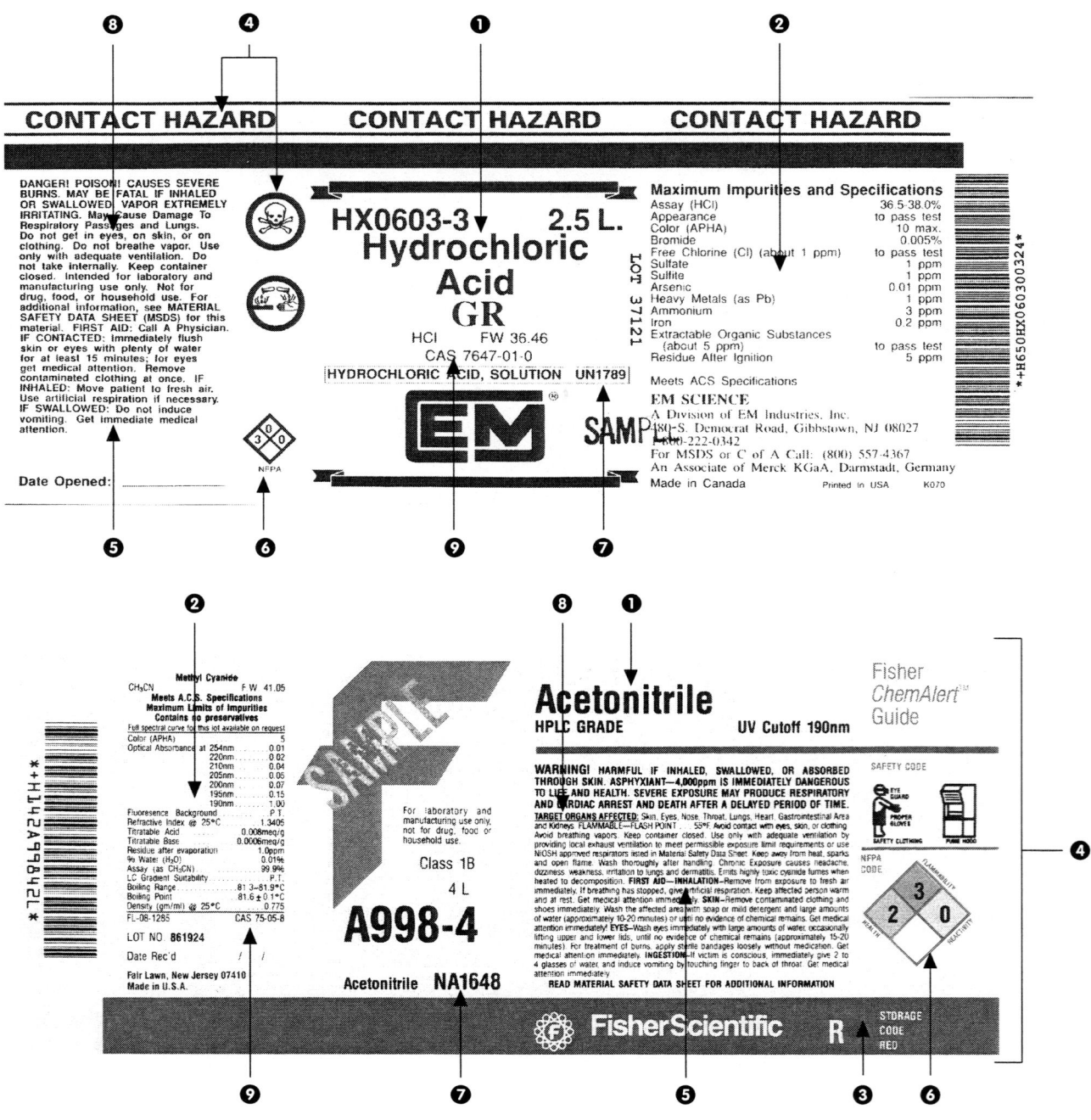

Figure 2B-1: Formats for Industrial Chemical Labels

Section 2B: Chemical Labels in the Laboratory and at Home

Exercise

In this Exercise, you will critically review industrial and consumer product labels for four chemicals. The objectives of this Exercise are to help you understand the role that labels play in keeping users informed, how industrial and consumer labels differ, and how being informed about a chemical can help prevent accidents.

Consumer product and industrial chemical labels are provided for 3% hydrogen peroxide (sold as a treatment for cuts), 5.25% sodium hypochlorite solution (household bleach), 70% isopropyl alcohol solution (rubbing alcohol), and toluene (sold as a paint thinner/ cleaner). The following checklists for labels are a compilation of the information presented in the Background. Evaluate each label according to the requirements of the appropriate checklist. Indicate whether the required information is present, understandable, and complete.

Consumer product label requirements (16 CFR 1500.121 and EPA)

The following are Consumer Product Safety Commission requirements unless otherwise noted.

- signal word DANGER, WARNING, or CAUTION

- signal word POISON present if required, along with signal word DANGER and the skull-and-crossbones symbol

- statement of principal hazard

- the common or chemical name of the hazardous substance(s)

- name and place of business of manufacturer, packer, distributor, or seller

- statement of additional hazards, if more than one hazard exists

- instructions for special handling or storage

- the statement "Keep out of reach of children" or its equivalent

- personal protective equipment needed for handling, such as gloves and goggles

- first-aid instructions for each hazard

- whether the chemical is a volatile organic compound (VOC) (required by EPA)

Industrial chemical product label requirements (ANSI Z129.1-1994 and OSHA 29 CFR 1910.1200)

The following are American National Standards Institute standards unless otherwise noted.

- legible symbols, pictures, and letters (required by OSHA)

- information presented in English (required by OSHA)

- identification of the chemical or its hazardous components (required by OSHA)

- statement of hazard(s) (required by OSHA)

- signal word DANGER, WARNING, or CAUTION

- precautionary measures, including protective equipment

- instructions in case of contact or exposure

- antidotes

- notes to physicians

- instructions in case of fire, spill, or leak

- instructions for container handling and storage

- manufacturer's name, business address, and phone number (required by OSHA)

- target organs

- MSDS availability and location

- POISON, along with skull-and-crossbones symbol (suggested for highly toxic chemicals—this does not replace signal word)

- CAS Registry Number, lot number, UN or NA number, specifications or lot analysis, size/weight of contents, and bar code

Answer the following questions:

1. What information is common to both consumer product and industrial chemical labels?

   The information common to both consumer product and industrial chemical labels are health hazard warnings, the name of the product, a list of ingredients, exposure treatment information, and proper storage conditions. Note that although manufacturer name and contact information are required for both by law, they are not shown in the consumer labels of this Exercise.

2. What information is provided on industrial chemical labels but is not present on consumer labels? Why would having this information be especially important in a workplace application?

   The information found on industrial chemical labels but not on consumer labels includes a referral to an MSDS and how to obtain one for that particular chemical, a fire diamond, a CAS Registry Number and a UN number, possibly pictograms of protective equipment, a list of primary and secondary hazards, the chemical's molecular formula, the chemical's intended use, and the possible reactions (such as oxidizing) with other chemicals. This information would be important in the workplace because the label explains the dangers associated with the chemical; identifies the means of protecting individuals; and covers emergency methods in case of accidental exposure, spills, or physical hazards. Note that the manufacturer name and contact information are required for consumer labels and need not be considered in this answer.

3. Explain how the consumer product labels meet the Consumer Product Safety Commission's requirement of presenting hazards "in conspicuous and legible type that contrasts by typography, layout, or color with other printed matter on the label."

   All these consumer product labels carry hazard warnings. Label designers used easy-to-read typefaces that contrast with their backgrounds. The typefaces of general hazard statements such as "Warning" or "Caution" are bold and sometimes in all capital letters and/or are larger than the surrounding text. Important directions are also bold and/or in all capital letters. Many directions are printed in plain text, so that the hazard statements will stand out and be easily found by the user. The use of boxes around warnings, especially for text written perpendicular to the label, helps the warning information stand out. Note that because these label examples were printed in black and white, color contrast in terms of the requirements cannot be discussed. However, it is important to consider readability for individuals who lack the ability to perceive colors.

4. Review the consumer labels provided (if possible, locate the actual material from the grocery store) and discuss strategies that are particularly effective in communicating safe handling information. Do the same for the industrial chemical labels. (In your analysis you should allow for the fact that these labels were scanned electronically, which caused loss of color and some print quality.)

   Discussions will differ but should include the readability and the legibility of the labels and hazard statements.

GENERAL STORAGE   GENERAL STORAGE   GENERAL STORAGE

WARNING! Weak Oxidizer. LABORATORY TESTS INDICATE MATERIAL MAY BE CARCINOGENIC. Do not get in eyes. Do not store near and avoid contact with combustible materials. Keep container closed and protected from light. Do not take internally. Intended for laboratory and manufacturing use only. Not for drug, food, or household use. For additional information, see MATERIAL SAFETY DATA SHEET (MSDS) for this material.

Date Opened: _____

HX0645-5   **1 L.**

Hydrogen Peroxide
GR

3% Solution

H_2O_2 FW 34.02

CAS 7722-84-1

LOT

SAMPLE

Maximum Impurities and Specifications		Typical Analysis
Assay	3.0% min.	3.5%
Free Acid (as H_2SO_4)	0.01%	0.001%
Chloride	0.001%	0.0002%
Phosphate	0.003%	0.0003%
Sulfate	0.005%	0.0005%
Arsenic	0.5 ppm	0.05 ppm
Nitrogen Compounds (as N)	0.005%	0.002%
Heavy Metals (as Pb)	1 ppm	0.1 ppm
Iron	5 ppm	0.5 ppm
Preservative	0.02%	0.01%
Residue After Evaporation	0.02%	0.006%

Hydrogen Peroxide CAS 7722-84-1
Water CAS 7732-18-5

EM SCIENCE
A Division of EM Industries, Inc.
480 S. Democrat Road, Gibbstown, NJ 08027
1-800-222-0342
For MSDS or C of A Call: (800) 557-4367
Associate of E. Merck, Darmstadt, Germany

Printed in USA   1251

SOLUTION OF
HYDROGEN PEROXIDE
3% H_2O_2 U.S.P.

TOPICAL ANTI-INFECTIVE

- For treatment of minor cuts and abrasions
- For use as a gargle or rinse

> DO NOT USE IF IMPRINTED SEAL ON CAP IS BROKEN OR REMOVED

INDICATIONS: For topical use as an antiseptic to help prevent infection in minor cuts, burns and abrasions, or to cleanse the mouth.

ACTIVE INGREDIENT: Hydrogen Peroxide 3%. Also Contains: 0.001% Phosphoric Acid as a stabilizer, and Purified Water.

STORAGE: Keep bottle tightly closed and at controlled room temperature 59° - 86°F (15° - 30°C). **DO NOT SHAKE BOTTLE.**

32 FL OZ (1 QT) 946 ml

DIRECTIONS: Apply locally to affected areas. To cleanse the mouth, dilute with an equal amount of water, swish around in mouth for at least one minute, then spit out, try to avoid swallowing this product.

WARNINGS: FOR EXTERNAL USE. Topically to the skin and mucous membranes. KEEP OUT OF EYES. Do not use on deep or puncture wounds or on serious burns. If redness, irritation, swelling or pain persists or increases or if infection occurs discontinue use and consult a physician. KEEP THIS AND ALL DRUGS OUT OF THE REACH OF CHILDREN. In case of accidental ingestion, seek professional assistance or contact a Poison Control Center immediately.

0   41250 56295   5

SAMPLE

L 871 45 6E F

HEALTH HAZARD   HEALTH HAZARD   HEALTH HAZARD

WARNING! VAPOR IRRITATING. HARMFUL IF INHALED. IRRITATING TO SKIN, EYES AND MUCOUS MEMBRANES. MAY CAUSE EYE INJURY. MAY BE HARMFUL IF SWALLOWED. Keep container closed. Do not breathe vapor. Do not get in eyes, on skin, or on clothing. Use with adequate ventilation. Do not take internally. Intended for laboratory and manufacturing use only. Not for drug, food, or household use. For additional information, see MATERIAL SAFETY DATA SHEET (MSDS) for this material. FIRST AID: Call A Physician. IF INHALED: Move patient to fresh air. Use artificial respiration if necessary. IF CONTACTED: Immediately flush skin or eyes with plenty of water for at least 15 minutes; for eyes get medical attention. IF SWALLOWED: Do not induce vomiting. Give milk or egg white beaten in water. Get immediate medical attention.

Date Opened: _____

UN1791

NFPA 2 0 0

SX0610-6   2.5 L.
Sodium Hypochlorite
Solution, Reagent
NaOCl   FW 74.45
CAS 7681-52-9

HYPOCHLORITE SOLUTION

 **SAMPLE**

Maximum Impurities and Specifications
Phosphate   5 ppm
Calcium   0.001%

Product will decompose with age; available chlorine at time of packing >6.0%.

Sodium Hypochlorite CAS 7681-52-9
Water CAS 7732-18-5

Store In A Cool Place Less Than 72°F

EM SCIENCE
Division of EM Industries, Inc.
480 S. Democrat Road, Gibbstown, NJ 08027
1-800-222-0342
For MSDS or C of A Call: (800) 557-4367
Associate of E. Merck, Darmstadt, Germany

Printed in USA   G280

SAMPLE

SPLASH RESISTANT with EASY OPEN SEAL

COMPARE TO CLOROX®

LEMON SCENT

WHITENS • CLEANS • BRIGHTENS • DEODORIZES • REMOVES STAINS

1 GALLON (3.79 L)

CAUTION: IRRITANT, DANGEROUS HARMFUL IF SWALLOWED Read Back Panel Carefully

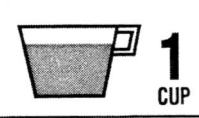
STANDARD WASHER
1 CUP

EXTRA LARGE WASHER
1½ CUPS

LAUNDRY USE

Sort laundry by color.
If uncertain about dye colorfastness, test fabric by applying one drop of a solution made of 1 tablespoon Meijer Lemon Scent Bleach plus 1/4 cup of water to hidden part of seam. Be sure to check all colors. After 1 minute, blot dry. No color change means the article can be safely bleached. Avoid bleaching wool, silk, mohair, leather, spandex and non-fast colors.

To hand wash or pretreat stains use 1/4 cup of bleach for every 1 gallon of cool water. Mix thoroughly before adding clothes.

Add Meijer Lemon Scent Bleach to dispenser if available. If not, add bleach and detergent with the wash water before the laundry is put in. For best results, dilute bleach with a quart of water and add to wash 5 minutes after start of wash cycles.

0   41250 64280   0

SAMPLE
Lemon Scent Bleach
brightens whites and removes stains

cleans and deodorizes

improves the effectiveness of detergents even in cold water

HOUSEHOLD HINTS
Use of Meijer Lemon Scent Bleach will leave your refrigerator smelling fresh and clean. Use it inside and out. Wash surfaces with a solution of 3/4 cup Meijer Lemon Scent Bleach per gallon of soapy water. Rinse and then air dry interior surfaces a few minutes before replacing food.

HOUSEHOLD USE

Toilet bowls. Flush toilet. Pour 1 cup of bleach into bowl. Brush. Let stand 10 minutes before flushing again.

Kitchen sinks. Cover stains with 1 gallon of water before adding 3/4 cup of Meijer Lemon Scent Bleach. Let stand 5 minutes before rinsing.

Bathtubs and showers, floors, vinyl, tile, woodwork and appliances. Clean with a solution of 3/4 cup of bleach per gallon of warm water. Let stand 5 minutes before rinsing.

100% Satisfaction Guaranteed

WARNING
Contains Sodium Hypochlorite 5.25%

Causes substantial but temporary eye injury. Avoid contact with eyes or on clothing. Harmful if swallowed. May irritate skin. For prolonged use, wear gloves.

Treatment: If in eyes rinse with plenty of water for 15 minutes. If swallowed, drink a glassful of water. Call physician in either case. If in contact with skin wash skin thoroughly with soap and water.

Add only to water, do not mix with acid or other household chemicals such as toilet bowl cleaners, ammonia or rust removers. To do so may release hazardous gases. Prolonged contact with metal may cause pitting or discoloration.

Storage and disposal: Store in a cool dry area, away from direct sunlight and heat to avoid deterioration. Do not reuse empty container; instead, rinse and put in trash collection.

*Clorox® is a registered trademark of The Clorox Company. Meijer Lemon Scent Bleach is not manufactured or distributed by The Clorox Company.

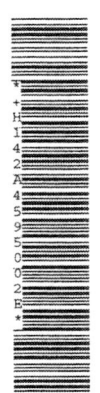

LOT NO. 977037

Isopropyl Alcohol   CAS 67-63-0
Water   CAS 7732-18-5

Fair Lawn, New Jersey 07410
Made in U.S.A.

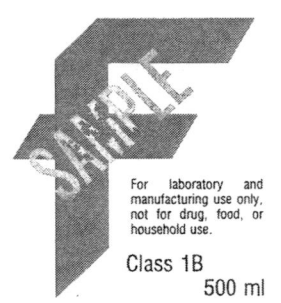

For laboratory and
manufacturing use only,
not for drug, food, or
household use.

Class 1B
500 ml

A459-500
Isopropanol Solution
UN1219
FL-01-0997

Isopropyl Alcohol
70% V/V

R

Warning! Flammable liquid. May cause eye and skin irritation. May cause respiratory and digestive tract irritation. May cause central nervous system depression. May cause kidney damage. For eye contact, flush with water and get medical aid. For skin contact, get medical aid if irritation occurs or persists. If ingested, give 2-4 cupfuls of milk or water and get medical aid. If inhaled, remove to fresh air and get medical aid.
IMPORTANT! Do not use this product until Material Safety Data Sheet has been read and understood.

MISE EN GARDE! Liquide inflammable. Risque de causer une irritation oculaire et cutanée. Risque de causer une irritation des voies respiratoires et digestives. Risque de causer une dépression du système nerveux central. Risque de causer une lésion des reins. En cas de contact oculaire, rincer avec de l'eau et obtenir des soins médicaux. En cas de contact cutané, si une irritation apparaît ou persiste, obtenir des soins médicaux. En cas d'ingestion, donner deux à quatre tasses de lait ou d'eau et obtenir des soins médicaux. En cas d'inhalation, transporter la victime à l'air frais et obtenir des soins médicaux.
IMPORTANT! Ne pas utiliser le produit avant d'avoir lu et compris la fiche signalétique.

 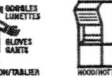

 Fisher Scientific

SAMPLE

70% Isopropyl
Rubbing Alcohol

RUBEFACIENT / TOPICAL ANTIMICROBIAL

(WARNING: FLAMMABLE, KEEP AWAY FROM FIRE OR FLAME.)

INDICATIONS: To decrease germs in minor cuts and scrapes. Helps relieve minor muscular aches due to overexertion.
DIRECTIONS: Apply to skin directly or with clean gauze, cotton or swab. For rubbing, apply liberally and rub with hands.
WARNINGS: FOR EXTERNAL USE ONLY. WILL PRODUCE SERIOUS GASTRIC DISTURBANCES IF TAKEN INTERNALLY. USE ONLY IN A WELL-VENTILATED AREA; FUMES MAY BE HARMFUL. KEEP OUT OF THE REACH OF CHILDREN. IN CASE OF ACCIDENTAL INGESTION, SEEK PROFESSIONAL ASSISTANCE OR CONTACT A POISON CONTROL CENTER IMMEDIATELY.
CAUTION: Do not apply to irritated skin, use in eyes or on mucous membranes. In case of deep or puncture wounds, consult doctor.
ACTIVE INGREDIENT: 70% Isopropyl Alcohol by volume.
Also contains: Water
NOTICE: Does not contain, nor is intended as substitute for grain or ethyl alcohol.

32 FL OZ (1 QT) 946 ml

L 810 45 6E FA

DANGER! FLAMMABLE LIQUID AND VAPOR. HARMFUL OR FATAL IF SWALLOWED. May Cause Damage To Central Nervous System, Liver, Kidneys and Lungs. VAPOR HARMFUL. CAUSES IRRITATION OF EYES, NOSE AND THROAT. Keep away from heat, sparks and flame. Use only with adequate ventilation. Do not breathe vapor. Do not get in eyes, on skin, or on clothing. Do not take internally. Intended for laboratory and manufacturing use only. Not for drug, food, or household use. Absorb spills with inert material, then place in a chemical waste container. Flush residual spill area with water. In case of fire, use water spray, foam, dry chemical or CO2. For additional information, see MATERIAL SAFETY DATA SHEET (MSDS) for this material. FIRST AID: Call A Physician. IF CONTACTED: Flush With Water. IF INHALED: Move patient to fresh air. Use artificial respiration if necessary. IF SWALLOWED: Call A Physician immediately ONLY induce vomiting at the instructions of a physician. Never give anything by mouth to an unconscious person. WARNING: This product contains a chemical(s) known to the State of California to cause birth defects or other reproductive harm.

Date Opened: _____

FLAMMABLE

TX0735-44   4 L.

Toluene

GR

$C_6H_5CH_3$ FW 92.14

CAS 108-88-3

TOLUENE   UN1294

LOT

Maximum Impurities and Specifications
Assay (GC)   min. 99.5%
Color (APHA)   10 max.
Sulfur Compounds (as S)   0.003%
Subs. Darkened By H_2SO_4 to pass test
Residue After Evaporation   0.001%
Water   0.03%

Meets ACS Specifications

SAMPLE

EM SCIENCE
A Division of EM Industries, Inc.
480 S. Democrat Road, Gibbstown, NJ 08027
1-800-222-0342
For MSDS or C of A Call: (800) 557-4367
An Associate of Merck KGaA, Darmstadt, Germany

Printed in USA   J323

77304

SAMPLE

Quality

TOLUENE

Photochemically Reactive

0 38514 77304 2

DANGER!
FLAMMABLE LIQUID AND VAPOR.
HARMFUL OR FATAL IF SWALLOWED.
VAPOR HARMFUL.
(Read Cautions on Back Panel)
871 grams VOC per liter.

32 FL. OZ. (ONE QUART) • 946 mL

CAUSES EYE, SKIN, NOSE and THROAT IRRITATION.

USE CAUTION WHEN OPENING! Contents may be under pressure. **KEEP AWAY FROM FACE.** Close container after use. *Do not transfer contents to unlabeled container for storage.* Do not store near heat, in areas of widely fluctuating temperatures, or display in windows.

KEEP OUT OF REACH OF CHILDREN.

VAPOR HARMFUL. MAY AFFECT THE BRAIN OR NERVOUS SYSTEM, CAUSING DIZZINESS, HEADACHE OR NAUSEA.

NOTICE: Repeated or prolonged solvent over-exposure may cause permanent brain/nervous system damage. Intentional misuse by deliberately concentrating and inhaling contents may be harmful or fatal.

VAPORS MAY CAUSE FLASH FIRE or ignite explosively. Keep away from heat, sparks and flames. DO NOT SMOKE. Extinguish all flames, pilot lights. Turn off stoves, heaters, electric motors, other sources of ignition during use and until <u>all</u> vapors are gone. Prevent vapor build-up by opening windows and doors for cross-ventilation and fresh air entry during application and drying. **USE ONLY WITH ADEQUATE VENTILATION.** Do not breathe vapors or spray mist. If you experience eye-watering, headaches or dizziness, increase fresh air, wear approved respiratory protection, or leave area. Avoid contact with eyes, skin, clothing. People having or suspected of having heart trouble, any pulmonary disorders, and women who are pregnant should consult physician before using. **FIRST AID:** If difficulty in breathing: leave area, obtain fresh air. If difficulty continues, get medical assistance immediately. *If swallowed, do not induce vomiting.* Call physician immediately. Eye contact: flush immediately with water. Get medical attention at once. Skin contact: wash thoroughly with soap and water.

DIRECTIONS FOR USE:

• **For thinning oil-base paints and varnishes:** add small amounts while stirring according to paint manufacturer's directions.
• **For cleaning painting equipment and removing spots from surfaces:** apply liberally <u>before</u> paint dries.
• **For removing wax, greas** ... cloth.
Do not use for cleaning asph ... e may result. <u>Do not use</u> to thin late...

WARNING!
THIS PRODUCT CONTAINS CHEMICALS KNOWN TO THE STATE OF CALIFORNIA TO CAUSE CANCER, BIRTH DEFECTS OR OTHER REPRODUCTIVE HARM.   71 - 75   4601

MADE IN U.S.A.

Section 2C: Looking at Label Codes and Symbols
Instructor Notes

The objective of this section is for students to become familiar with and to evaluate the advantages and disadvantages of common codes and symbols, including the National Fire Protection Association (NFPA) diamond, the Hazardous Material Information System (HMIS) bars, and common pictograms.

Student Pages

- Section 2C: Looking at Label Codes and Symbols—Background
- Section 2C: Looking at Label Codes and Symbols—Exercise

Possible Playout of the Section

After reading the Background, students discuss the advantages and disadvantages of using codes and symbols on chemical labels, including cultural or experiential barriers to understanding their meanings. Students then complete the Exercise.

Section 2C: Looking at Label Codes and Symbols

Background

Note: This section provides general recommendations for and information about chemical labels. However, you should always follow the procedures established by your organization's chemical hygiene plan.

Codes and symbols are often used on chemical labels because they offer consistency and quick recognition of relevant information. Codes and symbols can be effective communicators, but as *Precautionary Labels for Chemical Containers* states, "For communication to occur, however, all parties trying to communicate must agree on the form of communication and the meaning of the words and symbols used." Two frequently used forms of communicating hazard information are the National Fire Protection Association (NFPA) hazard diamond and the Hazardous Materials Information System (HMIS) bars. These are used in conjunction with additional text on the label. Pictograms and international symbols are also used frequently.

The NFPA Hazard Diamond

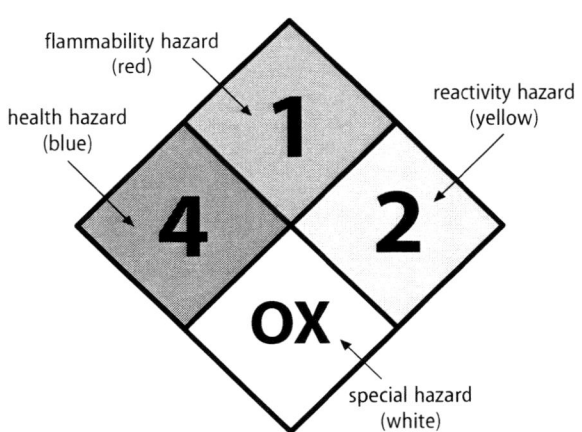

Figure 2C-1: Sample NFPA Diamond

The National Fire Protection Agency (NFPA) hazard diamond (adopted by NFPA in 1951) provides basic information to firefighters and other emergency personnel who must select firefighting tactics and emergency procedures. The NFPA hazard diamond is intended to address health, flammability, reactivity, and related hazards that may be presented by short-term exposure to a material during handling under conditions of fire, spill, or similar emergencies. It can be seen from a long distance, and the number codes can be read quickly. **The NFPA hazard diamond is NOT a good indicator for the hazards of using a chemical in a laboratory—the hazards during a fire can be and usually are dramatically different from the hazards of everyday laboratory use.**

The NFPA hazard diamond uses colors and numbers to convey hazard information. A large diamond is divided into four smaller diamonds, each a different color. (See Figure 2C-1.) The colors indicate the type of hazard, and the numbers indicate the severity. The meanings of these colors and numbers are described later in this section.

Because the hazard diamond was designed to provide basic information to emergency responders, it does not contain all of the information required by OSHA for chemical labels.

The NFPA Standard 704 specifically states that the NFPA hazard diamond does not apply to chronic exposure or non-emergency occupational exposure. Chemical labels using the NFPA hazard diamond must incorporate other information to meet OSHA regulations.

The Hazardous Materials Identification System

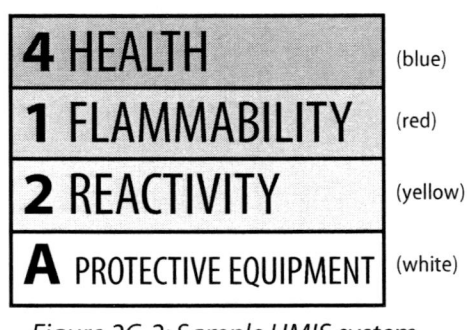

Figure 2C-2: Sample HMIS system
(Color version located on back cover.)

The Hazardous Materials Identification System (HMIS) was developed by the National Paint and Coatings Association (NPCA) to standardize the presentation of chemical information on labels. The HMIS system uses colored bars containing words and numbers. Color codes correspond to the hazards of a product; numeric ratings indicate the degree of hazard. (See Figure 2C-2.) The official HMIS system is licensed only for use by the Labelmaster company, but other label designs use an HMIS-type bar. Like the NFPA hazard diamond, HMIS is a standardized way of presenting hazard information. Unlike the NFPA hazard diamond, its primary purpose is not for emergency responders. HMIS includes alphabetical codes (A–Z) that designate the appropriate personal protective equipment (PPE) to wear while handling the material.

Understanding Color/Number Codes

In both the NFPA hazard diamond and the HMIS system, each color stands for a type of hazard. The colors blue, red, and yellow mean the same thing in both symbol systems: blue=health hazard, red=fire hazard, and yellow=reactivity hazard. White has a different meaning in each symbol system and is used in conjunction with code letters and symbols: white in the NFPA hazard diamond indicates a special hazard, and white in the HMIS system indicates the protective equipment required.

Examples of NFPA special hazard codes:
    OX = oxidizer
    W̶ = use no water
    ACID = acid
    COR = corrosive
    ALK = alkali

Examples of HMIS protective equipment codes:
    A = safety goggles
    B = safety goggles and gloves
    C = safety goggles, gloves, and synthetic apron
    D = safety goggles, gloves, synthetic apron, and face shield

In both the NFPA hazard diamond and the HMIS system, the hazard numbers range from 0 (minimum hazard) to 4 (severe hazard). The following list describes the general meanings of these numbers. The NFPA and HMIS methods each provide somewhat different wording for these ratings, but the basic meanings are the same.

Health Hazard (blue)—possibility of injury
    0 = no health hazard
    1 = could cause irritation
    2 = could cause temporary incapacitation
    3 = could cause serious temporary or irreversible injury
    4 = could cause death or irreversible injury

Flammability (red)—possibility of ignition
    0 = no fire hazard
    1 = must be preheated before ignition can occur
    2 = must be heated for ignition
    3 = flammable liquid or solid that can be readily ignited
    4 = flammable vapor or gas that burns readily

Reactivity (yellow)—possibility of reaction
    0 = stable material
    1 = may become unstable at high temperatures
    2 = readily capable of nonexplosive reaction
    3 = may detonate when exposed to heat or initiating source
    4 = readily capable of detonation or explosive reaction

Pictograms

Pictograms are often used on chemical labels to provide quickly recognized information about the type of hazard, target organs that may be damaged upon exposure, and/or protective equipment to be used. Pictograms are wordless drawings or symbols that represent a specific thing or idea. They provide the advantage of recognition by those who cannot read or those who do not read the language in which the label is written. However, pictograms may be so stylized that it is difficult to recognize what they represent, or they may depend on cultural norms. For example, bright yellow is used on warning signs in Japan, while in Europe, danger signs are outlined in red.

Section 2C: Looking at Label Codes and Symbols
Exercise

This Exercise contains a set of label codes and common pictograms. The purpose of the Exercise is to help you become familiar with and evaluate the advantages and disadvantages of common codes and symbols used in labeling chemicals.

Part 1: Label Codes

Each of the items below contains one error according to the conventions followed by the NFPA or HMIS label codes. Circle each error and briefly explain why it is wrong.

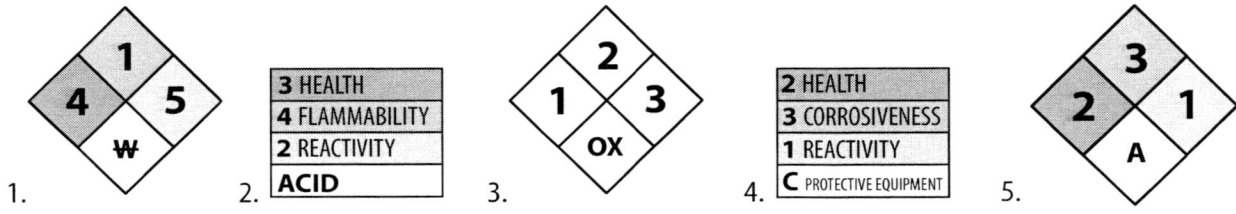

1. The number 5 on the right-hand side of the diamond is incorrect; the hazard numbers range from 0 to 4. 2. The bottom line incorrectly reads "Acid"; this line should list personal protective equipment. 3. The boxes in this diamond should be colored, not white. (Since this book is printed in black and white, the boxes should be different shades of gray.) 4. The word "corrosiveness" on the second line is incorrect; it should read "flammability." 5. The letter "A" at the bottom of the diamond is incorrect because "A" is not used by the NFPA. It is used by HMIS to represent goggles.

Part 2: Pictograms

Write down what you think each pictogram means, and indicate how effective this symbol is at conveying its message to you.

Chemical and Equipment Hazard Pictograms

16. 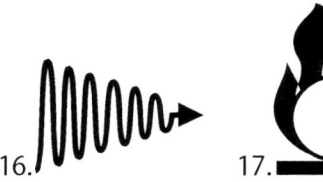 17.  18.

Personal Protective Equipment Pictograms

19. 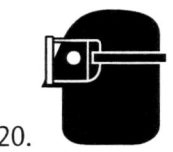 20.  21.  22.  23.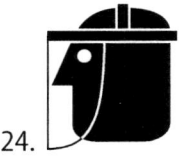

24.  25.  26.  27.  28.

Emergency Equipment Pictograms

29.  30.  31.

Pictogram key: 1. biohazard, 2. carcinogen, 3. corrosive, 4. explosive, 5. flammable, 6. hazard, 7. high voltage, 8. high voltage, 9. hot, 10. inhalation hazard, 11. microwave, 12. water reactive, 13. radiation, 14. poison, 15. static electricity, 16. ultraviolet hazard, 17. oxidizer, 18. highly reactive, 19. hearing protection, 20. goggles, 21. mask, 22. respirator, 23. safety glasses, 24. face shield, 25. apron, 26. foot protection, 27. gloves, 28. protective suit, 29. shower, 30. eyewash, 31. first aid.

Section 2D: Labeling Containers
Instructor Notes

The objective of this section is for students to understand the requirements for proper labeling of chemicals being stored in something other than their original container, being placed in a temporary container during an experiment, or being packed for shipment. Students use labels from commercially packaged products as references to practice making new labels for containers used in-house for storing and dispensing chemicals.

Student Pages

- Section 2D: Labeling Containers—Background
- Section 2D: Labeling Containers—Exercise

Possible Playout of the Section

Introduce the topic by asking students to describe situations in which chemicals are removed from their original containers and placed into secondary containers, such as storage bottles, squeeze bottles, or test tubes. You may wish to discuss safety concerns that might arise if these secondary containers were unlabeled.

Point out that one reason for using secondary containers is for shipping a chemical to someone outside of the laboratory. Have the class hypothesize about sending a sample to another laboratory for testing. Discuss the type of information that should be provided on the package to alert persons who might come into contact with the hazards.

For the Exercise, obtain original containers of reagent-grade acetone, nitric acid (concentrated), potassium permanganate, and sodium hydroxide. You may also wish to provide examples of some secondary labels.

Have students read the Background and complete the Exercise. As a class, you may wish to have the students share what hazard warnings they chose to put on the labels for acetone, concentrated nitric acid, potassium permanganate, and sodium hydroxide and explain their choices.

Section 2D: Labeling Containers

Background

Note: This section provides general recommendations for and information about chemical labels. However, you should always follow the procedures established by your organization's chemical hygiene plan.

The commercial chemicals you use in the laboratory are shipped in containers that are labeled by the manufacturer. Everyone who receives or uses chemicals should understand the labeling protocols used for chemicals that are shipped and heed warnings and precautions on these labels. In use, these chemicals may be transferred to different secondary containers for storage or dispensing (such as dispensing bottles, safety cans, and squeeze bottles) and to containers for immediate use (such as beakers, flasks, reaction vessels, and process equipment). OSHA does not regulate the labeling of chemicals in the laboratory once they have been removed from the original container. Labeling requirements for secondary containers should be stated in the chemical hygiene plan established by the organization.

Labeling Containers for Shipping

The U.S. Department of Transportation (DOT) strictly regulates the shipment of materials considered to be hazardous. These regulations can be found in the DOT Hazardous Materials Regulations, 49 CFR, Parts 100–185, and govern not only how the material is packaged but also how the outer package is marked and labeled and how the package is shipped. DOT considers materials to be hazardous if they pose a threat as a physical hazard (such as a fire or an explosion) or a health hazard (such as a poison or a corrosive).

Each material is assigned to a "packing group," which categorizes the material by how it should be packaged. The packing group is a DOT designation that indicates the degree of danger posed by the material. Packing Groups I, II, and III represent great, medium, and minor dangers, respectively. The extensive instructions for packaging each group are included in the DOT Hazardous Materials Regulations.

Some materials, such as compressed gases, do not have a Packing Group number designated because the level of their hazard is dependent upon the kind of cylinder the gas is contained in and the pressure of the gas in the cylinder.

After the proper packaging is selected and the material is packaged properly, the outside of the package must be marked and labeled. The "markings" are the shipping name, identification number, hazard class and division, specifications or UN/NA marks, and any required instructions and/or cautions. These markings are written on the outer package by the person preparing the package for shipment. "Labeling" means placing on the outside of the package a warning specific to the hazard class and/or handling precautions of the material. These package labels are generally preprinted (according to DOT specifications), and the person shipping the material merely selects the appropriate label(s) for the package. In the laboratory, you may be required to package chemicals for shipment, and you should know what the proper markings are and how to select the proper DOT labels.

- The shipping name is a descriptive name that tells the shipper what is in the container. The exact DOT shipping name must always be used.

- An identification number is assigned to each proper shipping name. Identification numbers preceded by the letters "UN" are associated with shipping names considered appropriate for international and domestic transportation. Identification numbers preceded by "NA" are associated with shipping names not recognized for international transportation, except to and from Canada.

- Any specifications, UN/NA marks, instructions, and/or cautions are dependent upon the materials being shipped. DOT has extensive lists and regulations governing what is required on the label or package of each material. Examples of DOT shipping information for some sample chemicals are shown in the following table:

Department of Transportation Hazardous Materials				
Shipping Name	Hazard Class and Division	Label Code	Identification Number	Packing Group
Charcoal	4.2	4.2	NA1361	III
Compressed Methane	2.1	2.1	UN1971	none
Hexanes	3	3	UN1208	II
Hydrogen Peroxide	5.1	5.1, 8	UN2014	II
Nitrogen Triiodide	Common mode of transportation forbidden; requires expert for special handling			
Lead Acetate	6.1	6.1	UN1616	III

- DOT assigns each regulated material to one of nine hazard classes that designate potential hazards posed by the material. The DOT Hazardous Materials Table (49 CFR 172.101) lists the hazard classes for all regulated materials. (This table can be accessed online at *hazmat.dot.gov.)* A material is assigned to the hazard class with the most serious applicable hazard. Since many materials have more than one hazard, when appropriate, DOT lists additional hazard classes on the Hazardous Materials Table. Appropriate DOT labels for all hazard classes listed on the Hazardous Materials Table must be applied to the outer package. The table on the next page lists the nine hazard classes, some with subdivisions.

Department of Transportation Hazard Classes and Symbols

Hazard Class/Division	DOT Package Label	Name of Class or Division	Hazard Class/Division	DOT Package Label	Name of Class or Division
1	EXPLOSIVE A	Explosives	5.1	OXIDIZER 5.1	Oxidizer
2.1	FLAMMABLE GAS	Flammable Gas	5.2	ORGANIC PEROXIDE 5.2	Organic Peroxide
2.2	NON-FLAMMABLE GAS	Non-flammable Gas	6.1	POISON	Poison
2.3	POISON GAS	Poison Gas	6.2	INFECTIOUS SUBSTANCE	Infectious Substance
3	FLAMMABLE LIQUID	Flammable and Combustible Liquid	7	RADIOACTIVE I	Radioactive
4.1	FLAMMABLE SOLID	Flammable Solid	8	CORROSIVE	Corrosive
4.2	SPONTANEOUSLY COMBUSTIBLE	Spontaneously Combustible	9	IRRITANT	Miscellaneous
4.3	DANGEROUS WHEN WET	Dangerous When Wet			

Labeling Secondary Laboratory Containers

It is frequently necessary to transfer chemicals from their primary, commercial container to secondary containers for temporary storage or immediate use. The contents of all secondary chemical containers should always be properly identified. *Prudent Practices in the Laboratory* states that "the overriding goal of prudent practice in the identification of laboratory chemicals is to avoid orphaned containers of unknown materials that may be expensive or dangerous to dispose of." If the container is so small that you cannot place a label on it—for example, a capillary tube or eyedropper—place it in a larger covered container and label the outer container. *Prudent Practices* also offers these guidelines for the labeling of laboratory chemical containers:

"The labels should be understandable to laboratory workers, members of well-trained emergency response teams, and others.... Chemical identification and hazard warning labels on outside packaging used for storing chemicals should include the following information:

- name, address, and telephone number of the chemical manufacturer, importer, or responsible party (including researcher),

- chemical identification and identity of hazard component(s), and

- appropriate hazard warnings.

Containers in immediate use, such as beakers and flasks, should, at a minimum, be labeled with the name of the chemical contents. Labeled materials transferred from primary (labeled) containers to secondary containers (for example, safety cans and squeeze bottles) should include chemical identification and synonyms, precautions, and first-aid information."

In addition to the above, good laboratory practice also includes writing the date the material was first stored and the initials of the preparer on the label of the secondary container.

Section 2D: Labeling Containers

Exercise

In this Exercise, you will use labels from the original containers of reagents to provide reference information as you make labels for secondary containers for storing and dispensing chemicals that might be used in a stockroom or at a laboratory bench.

For each of the following scenarios, write an appropriate label for the secondary storage or end-use container, and/or shipping package indicated.

1. You are filling a laboratory squeeze bottle with reagent-grade acetone.

    Include the name of the chemical ("Acetone"), the relevant hazard warning ("Flammable"), and the date and the initials of the preparer. Use a ballpoint pen or typed label covered with clear tape to avoid the problem of the acetone dissolving ink.

2. You are preparing 1 L of 3 M HNO_3 from concentrated. You will then store the 3 M HNO_3 until it is used.

    Include the name, the chemical formula, and molar concentration of the chemical ("3 M HNO_3; 3 M Nitric Acid"), as well as the date and the initials of the preparer.

3. You are making 2 L of 0.5 M $KMnO_4$ to be standardized and immediately used up for a titration by others in your group.

    Write the name, chemical formula, and molar concentration ("0.5 M $KMnO_4$; 0.5 M Potassium Permanganate), as well as the date and the initials of the preparer.

Chapter 3
Material Safety Data Sheets

At approximately 3:33 PM on June 20, 1994, an explosion and small fire occurred in a research laboratory when a mixture of 95% ethanol and sodium peroxide in a one-liter Erlenmeyer flask underwent an unexpected, uncontrolled reaction. The researcher received a small cut, second-degree burns to his hands, and minor chemical burns to his chest, legs, and face.... [T]he committee found that Material Safety Data Sheet (MSDS) information... indicate[s] that any reactions involving peroxidic materials in the presence of oxidizable organics have the potential for uncontrolled reactivity.

DOE, Lessons Learned Information Services Home Page

Chapter 3 Objectives

Pertinent Voluntary Industry Standards

- *VIS L2 Standards*—Recognize, apply, and respond appropriately to the hazard symbols and toxicology sections of MSDSs.

- *VIS (L2.02):06*—Read and interpret hazard data associated with chemicals that are presented in material safety data sheets (MSDSs) and other chemical data reference documents.

- *VIS (L2.02):10*—Demonstrate ability to convert chemical concentrations to different units so that comparison can be made with MSDS safe levels.

- *VIS (L2.02):11*—Access safety/environmental and health (S/H/E) regulations and data regarding chemicals using references such as CRC Press handbooks, the *Merck Index*, the *Chemical Technician's Ready Reference Handbook*, and MSDSs, as well as by conducting on-line searches.

- *VIS (L3.03):01*—Read and interpret standard operating procedures (SOPs) and material safety data sheets (MSDSs).

Goals of the Chapter

Students will learn

- how MSDSs can provide valuable safety information,

- what information is included in an MSDS,

- how the information on an MSDS can help protect them in the workplace, and

- the limitations of the MSDS as applied to laboratory-scale operations.

Why Your Students Need to Understand MSDSs

Material safety data sheets (MSDSs) provide detailed information beyond that found on labels. The passage of the OSHA Hazard Communication Standard of 1980 (often called the right-to-know law) called for the use of MSDSs as required documents for communicating with workers about hazardous chemicals. OSHA defines a chemical to be hazardous if it meets any of the following criteria:

- it is cancer-causing, toxic, corrosive, an irritant, a strong sensitizer, flammable, or reactive, and thus poses a threat to health;

- it is specifically listed under the Occupational Safety and Health Act, 29 CFR 1910, Section 1200; and/or

- it has an assigned threshold limit value (TLV) by the American Conference of Governmental and Industrial Hygienists (ACGIH).

Under OSHA, employees must have access to the MSDSs for the hazardous chemicals in the laboratory. Although OSHA does not require employers to maintain MSDSs for nonhazardous chemicals, many organizations choose to make available MSDSs for all chemicals they use due to continuing changes in regulations and the difficulty that often exists in predicting the exact hazard a chemical may pose. For example, some chemicals may be nonhazardous for most people, but for a few individuals they may be allergenic or act as "sensitizers," capable of generating an allergic reaction in a normally resistant individual after repeated exposure. In the workplace, graduates from your program will be expected to read the provided MSDSs for the chemicals they will be using and understand the information the MSDSs convey.

Right-to-know legislation is designed to provide workers with information to help them deal safely with occupational exposure to hazards. The objective of providing MSDSs is to concisely inform workers about the hazards of the materials they work with so that they can protect themselves and others and so they can respond properly to emergency situations. However, MSDSs have in practice been designed more for health and safety professionals, emergency responders such as firefighters, and medical personnel. Therefore, MSDSs often do a poor job of conveying information to laboratory practitioners, including chemists and chemical technicians.

Prudent Practices discusses significant limitations of MSDSs as applied to laboratory settings, explaining that MSDSs must describe hazard warnings and controls for work on a wide scale, even though chemicals could be used in microscale operations as well as manufacturing operations. As a result, control measures such as use of personal protective equipment or spill cleanup stated in an MSDS may be unnecessary or inappropriate in a laboratory setting. The practitioner often has the difficult task of judging the relevance of the information presented in the MSDS. In addition, according to *Prudent Practices*, "Many MSDSs comprehensively list all conceivable health hazards associated with a substance without differentiating which are most significant and which are most likely to be encountered. This can make it difficult for laboratory workers to distinguish highly hazardous materials from moderately hazardous and relatively harmless ones."

J. Kaufman of the Laboratory Safety Institute discusses additional limitations of MSDSs in the *Laboratory Safety Newsletter:* "My observation has been that sheets tend to straddle the right-to-know issue. Some appear to be written from a marketing standpoint, as if any admission of a product's possible hazard would make it unsalable. . . . From the other side, some sheets give so many dire warnings that the average non-technically trained user is afraid to even read the label, let alone use the chemical. Sure, typing correction fluid has a solvent in it, but I doubt that the average user needs gloves or a respirator." (Kaufman, *Laboratory Safety Newsletter,* Vol. 2, #4) To add to the potential problems with MSDSs, the quality, readability, and completeness of MSDSs vary widely depending on the effort put forth by the producers.

Overview of the Chapter

This chapter contains five sections dealing with MSDSs. General background information and complete instructions for the exercises are provided. The background pages are intended only to provide basic background information on MSDSs for your students. We recommend that you extend this information by having students find and read the pertinent sections of the books in the safety reference library (described in Chapter 1). You may wish to have students briefly report on their findings during the monthly safety meetings.

MSDSs are also included as part of the student section with the permission of the chemical vendors who developed them. Each MSDS or excerpt cites the original source. These MSDSs were current at time of publication; you may choose to replace them with more current ones from time to time. **The MSDSs included are intended for instructional purposes only and are not intended to provide specific safety instructions for students in your laboratory.**

Section 3A: Getting to Know MSDSs

This section introduces students to MSDSs and the regulations governing their use. To initiate interest in the topic, students read and discuss Laboratory Incident Reports concerning situations in which heeding information provided in an MSDS could have prevented an accident. These examples serve to get students thinking about the importance of MSDSs in the academic and industrial laboratory.

Section 3B: Exploring MSDSs

The objective of this section is for students to become familiar with the content and scope of MSDSs by comparing the industrial labels presented in Chapter 2, Section 2B, with MSDSs for the same chemicals.

Section 3C: Understanding Health Hazards

In this section, students become familiar with the health hazard information provided on MSDSs. They review the MSDS health data for a number of solvents, select those posing the least health hazards, and explain their reasoning.

Section 3D: Finding MSDSs on the Web

The objective of this section is for students to use web-browsing software to locate MSDSs on the Web and apply the information from the MSDSs to answer questions about certain chemicals.

Section 3E: Evaluating MSDSs

To become familiar with critiquing MSDSs, students will learn to assess MSDSs for completeness using a set of evaluation questions.

References for Chapter 3

Annual Report on Carcinogens. National Toxicology Program: Research Triangle Park, NC, 1999.

American Cancer Society. Cancer Statistics Webpage. www.cancer.org/statistics/index.html (accessed Sept 16, 1999).

American Conference of Governmental and Industrial Hygienists (ACGIH). *Threshold Limit Values for Chemical Substances and Physical Agents and Biological Exposure Indices.* 1999.

"Do You QC Your MSDSs?," *Laboratory Safety Newsletter, 2* (4). Kaufman & Associates: Natick, MA, 1995.

Epidemiology and Risk Assessment; Gordis, L., Ed.; Oxford University: New York, 1988.

Flinn Chemical and Biological Catalog Reference Manual 1998; Flinn Scientific: Batavia, IL, 1998.

Hazard Communication Standard. *Code of Federal Regulations,* 29 CFR 1910.1200, 1999. (This is available free online at www.osha.gov.)

Hazardous Industrial Chemicals—Material Safety Data Sheets—Preparation. *American National Standards Institute Standards,* ANSI Z400.1-1993. American National Standards Institute: Washington, DC.

Hofstader, R.; Chapman, K. *Foundations for Excellence in the Chemical Process Industries: Voluntary Industry Standards for Chemical Process Industries Technical Workers;* American Chemical Society: Washington, DC, 1997.

Improving Safety in the Chemical Laboratory: A Practical Guide, 2nd ed.; Young, J.A., Ed.; John Wiley & Sons: New York, 1991.

International Agency for Research on Cancer (IARC). Lists of IARC Evaluations. IARC monograms Website. 193.51.164.11/monoeval/grlist.html (accessed Sept 17, 1999).

Learning By Accident, Vol. 2; Mojtabai, F.; Kaufman, J., Eds.; Laboratory Safety Institute: Natick, MA, 2000.

Occupational Exposure to Hazardous Chemicals in Laboratories Standard. *Code of Federal Regulations,* 29 CFR 1910.1450, 1999. (This is available free online at www.osha.gov.)

National Institute of Health's National Toxicology Program Chemical Health and Safety Database. ntp-server.niehs.nih.gov/ (accessed Feb 29, 2000)

Prudent Practices in the Laboratory: Handling and Disposal of Chemicals; National Academy: Washington, DC, 1995.

U.S. Department of Energy. Lessons Learned Information Services Homepage. tis.eh.doe.gov/ll/ (accessed Feb 7, 2000).

Section 3A: Getting to Know MSDSs
Instructor Notes

This section introduces students to MSDSs and the regulations governing their use. To initiate interest in the topic, students read and discuss Laboratory Incident Reports concerning situations in which heeding information provided in an MSDS could have prevented an accident. These examples serve to get students thinking about the importance of MSDSs in the academic and industrial laboratory.

Student Pages

- Section 3A: Getting to Know MSDSs—Laboratory Incident Reports
- Section 3A: Getting to Know MSDSs—Background

Possible Playout of the Section

Students read both the Laboratory Incident Reports and Background and discuss the following questions:

1. What are similarities and differences between the incidents described in the Laboratory Incident Reports?

2. What role could MSDSs have played in preventing each of the accidents described in the Laboratory Incident Reports?

3. Explain the importance of MSDSs in the workplace.

4. If a colleague told you that reading MSDSs wasn't worth the trouble, how would you respond?

5. What are the legal implications of not having MSDSs available?

Section 3A: Getting to Know MSDSs

Laboratory Incident Reports

Incident Descriptions

At approximately 3:33 PM on June 20, 1994, an explosion and small fire occurred in a research laboratory when a mixture of 95% ethanol and sodium peroxide in a one-liter Erlenmeyer flask underwent an unexpected, uncontrolled reaction. The researcher received a small cut, second-degree burns to his hands, and minor chemical burns to his chest, legs, and face.... [T]he committee found that Material Safety Data Sheet (MSDS) information... indicate[s] that any reactions involving peroxidic materials in the presence of oxidizable organics have the potential for uncontrolled reactivity.

DOE, Lessons Learned Information Services Home Page

When I was a graduate student, my laboratory bench partner made only infrequent visits to our lab. Unwashed, unlabeled glassware tended to accumulate on her side of the bench. One day she appeared, intent upon cleaning her area. A 500-mL round-bottomed flask on the bench contained a reasonable amount of white solid. We later found that the solid had been formed by mixing solutions of one of the borazines and silver perchlorate and allowing the solvent to evaporate. My partner was not wearing eye protection when she placed a small amount of the white material on a metal spatula and immersed it in a stream of tap water at the sink. The white material exploded violently. Next, my partner directed a fast-flowing stream of tap water into the flask itself. An immediate detonation followed. I was standing, back turned, about six feet away and was deafened for several minutes. Flying glass shards penetrated all the fluorescent light tubes within a twenty-foot radius, darkening the room. The force of the explosion blew the tops off three nearby gallon bottles of organic solvent at the liquid line, setting their contents—benzene, acetone and isopropanol—on fire.

The ambulance arrived within ten minutes. My partner suffered severe lacerations, particularly in the arm and hand which had held the flask. She also received a number of glass shards, too small to be removed, into one eye.

Learning by Accident, Vol. 2, page 72

In a college organic chemistry lab during the mid-1970s, after the instructor left the room, a student began doing his own experiment. Claiming to know what he was doing, he combined chemicals to form a very unstable liquid compound, but he inadvertently spilled it. Attempts to mop up the spill produced explosions whenever the chemical came into contact with cleaning materials. Chemical-soaked towels thrown into the trash can caused a small explosion, which was large enough to cause the instructor to return.

Learning by Accident, Vol. 2, page 62

Section 3A: Getting to Know MSDSs
Background

Note: This section provides general recommendations for and information about MSDSs. However, you should always follow the procedures established by your organization's chemical hygiene plan.

"Look! I found the correct solvent application instructions!"

In an attempt to protect employees who are exposed to chemicals in the workplace, the United States Government created the Hazard Communication Standard, which requires that all firms manufacturing and/or distributing chemicals for use in the United States provide material safety data sheets (MSDSs) for those chemicals and distribute them to their customers. The law further requires that employers must make available to their employees an MSDS for every hazardous chemical that is present in their workplace.

OSHA defines a chemical to be hazardous if it meets any of the following criteria:

- it is cancer-causing, toxic, corrosive, an irritant, a strong sensitizer, flammable, or reactive, and thus poses a threat to health;

- it is specifically listed under the Occupational Safety and Health Act, 29 CFR 1910, Section 1200; and/or

- it has an assigned threshold limit value (TLV) by the American Conference of Governmental Hygienists.

Employers must also provide employees with the information and training necessary to read and understand MSDSs so that those employees can do what is necessary to protect themselves from the hazards associated with exposure to these chemicals.

What Is an MSDS?

An MSDS provides concise information about the hazards of a particular chemical so that persons handling or storing these materials can protect themselves and their co-workers as well as respond to emergency situations. Knowing the information on the MSDS for each chemical you work with and how to interpret this information can help to reduce human and environmental health risks, minimize accidents, and save money. An MSDS summarizes facts from many sources. Training, knowledge, and understanding of the technical data on an MSDS will provide you with the skills, knowledge, and good judgment to safely deal with your occupational exposure to hazards. Specific legal requirements for an MSDS in the

U.S. are addressed in OSHA's Hazard Communication Standard. According to OSHA's Hazard Communication Standard (as of July 1999), each MSDS must include:

- the material's identity, including its chemical and common names as shown on the label (example, brand name: Clorox®; chemical name: sodium hypochlorite; common name: bleach), the identity used shall permit cross-referencing to be made among the required lists of hazardous chemicals, the label, and the MSDS;

- hazardous ingredients that occur in the mixture in parts as small as 1%;

- cancer-causing ingredients that occur in the mixture in parts as small as 0.1%;

- list of properties of the material (reactive, flammable, explosive, unstable, etc.);

- list of acute and chronic health hazards;

- precautions and safety equipment;

- emergency and first aid procedures;

- specific firefighting information;

- procedures for cleaning up spills and leaks;

- precautions for safe handling and use, including personal hygiene; and

- identity of the organization responsible for creating the MSDS, date of issue, and emergency phone number.

The MSDS Format

Although all MSDSs must provide the information just listed, no standard format is required. A few years ago the Chemical Manufacturers Association (CMA) began working on a standard to develop consistent and understandable MSDSs that would be easily recognized in the U.S., Canada, and Europe. This standard, also accepted by ANSI (American National Standards Institute), does not fulfill every nation's legal requirements, but it does provide consistent and useful information to a variety of audiences. OSHA's requirements are covered in the first 10 of 16 total sections of the CMA/ANSI format, but it's not yet known whether OSHA will adopt the format. (ANSI standards, although well respected and followed by most industries, are recommendations, not laws.) *You will become familiar with the contents and format of MSDSs in Section 3B: Exploring MSDSs.*

Understanding Health Hazards

One of the most important reasons for reading MSDSs is to understand the potential health hazards of the chemicals you are working with. The health hazard information on MSDSs contains many unfamiliar abbreviations, and most people find it challenging to read and understand the significance of the information provided. After all, "lethal concentration 50," "lethal dose 50," and "lethal dose low" all sound bad, but what is the difference between these terms, and what do they mean to you as you contemplate using a particular chemical for a procedure? *You will practice reading and using the health hazard information on MSDSs in Section 3C: Understanding Health Hazards.*

Obtaining MSDSs for the Chemicals You Work With

In the workplace, your employer is required to make available to you the MSDSs for all hazardous chemicals you may be exposed to. The MSDSs may be provided to you directly, may be available for your reference in a central location, or may be available on the computer through a company database. Additionally, you can find MSDSs for many thousands of chemicals via the Web. *You will practice locating MSDSs on the World Wide Web in Section 3D: Finding MSDSs on the Web.*

Limitations of MSDSs

MSDSs were not originally developed for use by chemists and chemical technicians. Although they are valuable resources, the usefulness of MSDSs in a laboratory setting is limited in several ways (as discussed in later sections); and because they are developed by many different sources and lack a required format, their quality varies. You can supplement the information provided in MSDSs by using references such as the CRC handbooks, *Merck Index, Chemical Technicians' Handbook,* and *Prudent Practices in the Laboratory. You will practice evaluating MSDSs and using additional resources in Section 3E: Evaluating MSDSs.*

Section 3B: Exploring MSDSs
Instructor Notes

The objective of this section is for students to become familiar with the content and scope of MSDSs by comparing the industrial labels presented in Chapter 2, Section 2B, with MSDSs for the same chemicals.

Student Pages

- Section 3B: Exploring MSDSs—Background
- Section 3B: Exploring MSDSs—Exercise

Possible Playout of the Section

By reading the Background and completing the Exercise, students learn the types of information provided by MSDSs, how to use them in an industrial operation or chemical experiment, and how to efficiently scan them for pertinent safety information, including information about human health, ecological impact, and regulations. Although MSDSs are a good source of information about a chemical, it should be pointed out that they are not the only source of information. MSDSs are used in conjunction with chemical hygiene plans, standard operating procedures, and hands-on training received in the laboratory.

After the students have had time to explore the MSDSs for 3% H_2O_2 and toluene (which are provided following the Exercise) and complete the Exercise, discuss with the students their answers to questions 3 and 4. You may also wish to discuss the MSDSs in general, including content, format, and layout issues.

Section 3B: Exploring MSDSs

Background

Note: This section provides general recommendations for and information about MSDSs. However, you should always follow the procedures established by your organization's chemical hygiene plan.

An MSDS for a single chemical can be quite long, perhaps 12 pages or more. The MSDS is filled with numbers and abbreviations, many of them unfamiliar to inexperienced readers. While reading and understanding a document like this may seem daunting, knowing what information to expect and where to find it makes the task easier. The voluntary American National Standards Institute (ANSI) format helps by ensuring that information always appears in the same 16 sections, organized around four key issues that address what people need to know about working with chemicals. The four key issues and the sections related to each are described below. The first 10 sections cover the information required by OSHA standards. Sections 11–16 provide additional information.

What is the material and what do I need to know immediately in an emergency?

Section 1: Chemical product and company identification

- identifies material; should be the same name as the label
- identifies the supplier of the MSDS
- identifies a source for more information

Section 2: Composition/information on ingredients

- lists the OSHA hazardous components, chemical and common name if appropriate
- lists corresponding CAS Registry Numbers
- may also list significant nonhazardous components
- may also include additional information about components (for example, exposure guidelines)

Section 3: Hazards identification, including emergency overview

- may provide material description (for example, form, color, or odor)
- may provide emergency overview
- provides information on the potential adverse health effects and symptoms that might result from reasonably foreseeable use and misuse of the material

What should I do if a hazardous situation occurs?

Section 4: First-aid measures

- provides instructions to be followed if accidental exposure requires immediate treatment
- may also include instructions to medical professionals

Section 5: Firefighting measures

- provides basic firefighting guidance, including appropriate extinguishing media
- describes other fire and explosive properties useful for avoiding and fighting fires involving the material, such as flash point or explosive limits
- includes NFPA hazard diamond code

Section 6: Accidental release measures

- describes actions to be taken to minimize the adverse effects of an accidental spill, leak, or release of the material

How can I prevent hazardous situations from occurring?

Section 7: Handling and storage

- provides information on safe handling and storage

Section 8: Exposure controls/personal protection

- provides information on practices and/or equipment that are useful in minimizing worker exposure
- may also include exposure guidelines
- provides guidance on personal protective equipment

Section 9: Physical and chemical properties

- provides additional data that can be used to help characterize the material and design safe work practices

Section 10: Stability and reactivity

- describes the conditions to be avoided or other materials that may cause a reaction that would change the intrinsic stability of the material

Is there any other useful information about this material?

Section 11: Toxicological information

- may provide background toxicological information on the material, its compounds, or both
- usually includes LD_{50} for oral toxicity and LC_{50} for inhalation toxicity; also includes acute or long-term effects and target organs

Section 12: Ecological information

- may provide information on the effects the material may have on plants or animals and information on the material's environmental fate if the material is accidentally released into the environment

Section 13: Disposal considerations

- may provide information that is useful in determining appropriate disposal measures

Section 14: Transport information

- may provide basic shipping classification information
- includes the UN (NA) number and specific DOT information

Section 15: Regulatory information

- may provide any additional information on regulations affecting the material, such as the Resource Conservation and Recovery Act (RCRA)

Section 16: Other information

- may provide any additional information

Section 3B: Exploring MSDSs

Exercise

Knowing how to read and understand MSDSs is a critical part of working in the chemical laboratory or chemical technology field. Reading MSDSs can be intimidating at first, because they are often long and filled with unfamiliar information. However, a familiarity with the content and layout of MSDSs is vital and might even save your life or a co-worker's life someday.

This Exercise provides a familiar frame of reference by using MSDSs for the same chemicals (3% hydrogen peroxide and toluene) whose labels were presented in the Exercise in Chapter 2, Section 2B.

Look over the MSDSs (provided on the following pages) and the corresponding commercial labels (from Exercise 2B). Locate each piece of information shown on the label on the corresponding MSDS. Use a highlighter to mark information on the MSDS that is also shown on the label. Note that information on a particular topic may appear in more than one place on the MSDS. Compare the label's and the MSDS's presentation of similar information. How are they similar? Different? For example, where a label might use the word "flammable" to describe a chemical, an MSDS might provide the flash point and other data pertaining to flammability.

Answers will depend on the information highlighted on the labels that corresponds with the information in the MSDSs.

Look over the non-highlighted information on the MSDS. What kinds of information does the MSDS contain that the label does not? What, if any, significant differences were observed between the types of information found on the labels and MSDSs?

The information in the MSDSs is typically more specific than the information on the labels. The MSDSs usually contain information on firefighting measures, accidental release measures, handling and storage, exposure controls and personal protection, physical and chemical properties of the chemical, stability and reactivity, toxicological information, ecological information, disposal considerations, transport information, regulatory information, and additional information (stating the company does not assume liability for damages and such).

MATERIAL SAFETY DATA SHEET—3% Hydrogen Peroxide

SECTION 1—Chemical Product and Company Identification

MSDS Name: Hydrogen Peroxide 3%
Synonyms: Hydrogen dioxide, hydroperoxide
Company ID: Fisher Scientific, 1 Reagent Lane, Fairlawn, NJ 07410
For information call: 201-796-7100
Emergency Number: 201-796-7100
For CHEMTREC assistance call: 800-424-9300
For International CHEMTREC assistance call: 703-527-3887

SECTION 2—Composition, Information on Ingredients

CAS#	Chemical Name	%	EINECS#
7722-84-1	Hydrogen peroxide	3.0	231-765-0
7732-18-5	Water	97	231-791-2

Hazard Symbols: XI
Risk Phrases: 36/38

SECTION 3—Hazards Identification

Emergency Overview

Appearance: clear, colorless
CAUTION! May cause eye and skin irritation. May cause respiratory and digestive tract irritation.
Target Organs: Respiratory system

Potential Health Effects

Eye:	Produces irritation, characterized by a burning sensation, redness, tearing, inflammation, and possible corneal injury. Vapors may cause eye irritation.
Skin:	May cause skin irritation.
Ingestion:	May cause irritation of the digestive tract.
Inhalation:	May cause respiratory tract irritation. Irritation may lead to chemical pneumonitis and pulmonary edema.
Chronic:	Not available.

SECTION 4—First Aid Measures

Eyes:	Immediately flush eyes with plenty of water for at least 15 minutes, occasionally lifting the upper and lower lids. Get medical aid immediately. Do NOT allow victim to rub or keep eyes closed.
Skin:	Flush skin with plenty of soap and water for at least 15 minutes while removing contaminated clothing and shoes. Get medical aid if irritation develops or persists.
Ingestion:	Do NOT induce vomiting. If victim is conscious and alert, give 2–4 cupfuls of milk or water. Get medical aid immediately. Call a poison control center.
Inhalation:	Remove from exposure to fresh air immediately. If not breathing, give artificial respiration. If breathing is difficult, give oxygen. Get medical aid.
Notes to Physician:	Treat symptomatically and supportively.
Antidote:	No specific antidote exists.

SECTION 5—Firefighting Measures

General Information:	Use water spray to keep fire-exposed containers cool. Wear appropriate protective clothing to prevent contact with skin and eyes. Wear a self-contained breathing apparatus (SCBA) to prevent contact with thermal decomposition products. Substance is noncombustible.
Extinguishing Media:	Use water only! Do NOT use carbon dioxide. Do NOT use dry chemical.
Autoignition Temperature:	Noncombustible
Flash Point:	Noncombustible
NFPA Rating:	Not published
Explosion Limits	Lower: 40
	Upper: 100

SECTION 6—Accidental Release Measures

General Information: Use proper personal protective equipment as indicated in Section 8.

Spills/Leaks: Use water spray to disperse the gas/vapor. Absorb spill using an absorbent, noncombustible material such as earth, sand, or vermiculite.

SECTION 7—Handling and Storage

Handling: Wash thoroughly after handling. Remove contaminated clothing and wash before reuse. Use with adequate ventilation. Do not get on skin or in eyes. Do not ingest or inhale.

Storage: Keep away from heat and flame. Do not store in direct sunlight. Store in a cool, dry, well-ventilated area away from incompatible substances.

SECTION 8—Exposure Controls, Personal Protection

Engineering Controls: Use adequate general or local exhaust ventilation to keep airborne concentrations below the permissible exposure limits. Exercise care in cleaning all equipment before use with substance to prevent contamination.

Exposure Limits:

Chemical Name	ACGIH	NIOSH	OSHA—Final PELs
Hydrogen peroxide	1 ppm	1 ppm TWA	1 ppm TWA
	1.4 mg/m^3 TWA	75mg/m^3	1.4 ppm IDLH
Water	none listed	none listed	none listed

OSHA Vacated PELs

Hydrogen peroxide: 1 ppm TWA; 1.4 mg/m^3 TWA

Water: No OSHA Vacated PELs are listed for this chemical.

Personal Protective Equipment

Eyes: Wear appropriate protective eyeglasses or chemical safety goggles as described by OSHA's eye and face protection regulations in 29 CFR 1910.133.

Skin: Wear appropriate protective gloves to prevent skin exposure.

Clothing: Wear appropriate protective clothing to prevent skin exposure.

Respirators: Follow the OSHA respirator regulations found in 29 CFR 1910.134. Always use a NIOSH-approved respirator when necessary.

SECTION 9—Physical and Chemical Properties

Physical State:	Liquid
Appearance:	Clear, colorless
Odor:	Odorless
pH:	5.0–6.0 (1% sol)
Vapor Pressure:	23 mm Hg @30°C
Vapor Density:	Not available.
Evaporation Rate:	>1.0 (Butyl acetate=1)
Viscosity:	1.25 cP
Boiling Point:	212°F
Freezing/Melting Point:	32°F
Decomposition Temperature:	Not available
Solubility:	Completely soluble in water
Specific Gravity/Density:	1.0 (3%)
Molecular Formula:	H_2O_2
Molecular Weight:	34.0128

SECTION 10—Stability and Reactivity

Chemical Stability: Decomposes slowly to release oxygen

Conditions to Avoid: Incompatible materials, light, excess heat

Incompatibilities with Other Materials: Acetic acid, acetic anhydride, acetone, alcohols + sulfuric acid, ammonia and its carbonates, antimony trisulfide, arsenic trisulfide, brass, bronze, t-butyl alcohol, cellulose, charcoal, chlorine + potassium hydroxide, chlorosulfonic acid, chromium, copper, cupric sulfide, ethyl alcohol, ferrous sulfide, formic acid + organic matter, gold, hydrazine, hydrogen selenide, iron, ketones + nitric acid, lead, lead dioxide, lead monoxide, lead sulfide, lime water, magnesium, manganese, manganese dioxide, mercuric oxide, mercurous oxide, molybdenum disulfide, nitric acid, organic matter, 1-phenyl-2-methylpropyl alcohol + sulfuric acid, platinum, potassium, potassium permanganate, silver, sodium, sodium iodate, thiodiglycol, unsymmetrical dimethylhydrazine

Hazardous Decomposition Products: Oxygen, hydrogen gas

Hazardous Polymerization: Has not been reported

SECTION 11—Toxicological Information

RTECS#

CAS# 7722-84-1:	MX0899000 MX0900000
CAS# 7732-18-5:	ZC0110000

LD_{50}/LC_{50}

CAS# 7722-84-1:	Inhalation, rat: LC_{50} = 2 mg/m^3/4H; Oral, mouse: LD_{50} = 2 mg/kg; Skin, rat: LD_{50} = 4060 mg/kg.
CAS# 7732-18-5:	Oral, rat: LD_{50} = >90 mL/kg.
Carcinogenicity:	Hydrogen peroxide: ACGIH: A3—Animal Carcinogen, IARC: Group 3 carcinogen
	Water: Not listed by ACGIH, IARC, NIOSH, NTP, or OSHA
Epidemiology:	No information available
Teratogenicity:	No information available
Reproductive Effects:	No information available
Neurotoxicity:	No information available
Mutagenicity:	No information available
Other Studies:	None

SECTION 12—Ecological Information

Ecotoxicity:	Fingerling trout, >40 ppm/48H is toxic. Fathead minnow LC_{50} = 22–35 mg/L/96H
Environmental Fate:	No information reported
Physical/Chemical:	No information available
Other:	None

SECTION 13—Disposal Considerations

Dispose of in a manner consistent with federal, state, and local regulations.

RCRA D-Series Maximum Concentration of Contaminants: None listed

RCRA D-Series Chronic Toxicity Reference Levels: None listed

RCRA F-Series: None listed

RCRA P-Series: None listed

RCRA U-Series: None listed

Not listed as a material banned from land disposal according to RCRA

SECTION 14—Transport Information

US DOT:	No information available
IMO:	No information available
IATA:	No information available
RID/ADR:	No information available
Canadian TDG:	No information available

SECTION 15—Regulatory Information

US Federal

TSCA

CAS# 7722-84-1 is listed on the TSCA inventory.

CAS# 7732-18-5 is listed on the TSCA inventory.

Health & Safety Reporting List: None of the chemicals are on the Health & Safety Reporting List.

Chemical Test Rules: None of the chemicals in this product are under a Chemical Test Rule.

Section 12b: None of the chemicals are listed under TSCA Section 12b.

TSCA Significant New Use Rule: None of the chemicals in this material have a SNUR under TSCA.

SARA

Section 302 (RQ):	None of the chemicals in this material have an RQ.
Section 302 (TPQ):	CAS# 7722-84-1: concentration > 52%: TPQ = 1,000 pounds; RQ = 1,000 pounds
SARA Codes:	CAS # 7722-84-1: acute, flammable
Section 313:	No chemicals are reportable under Section 313.

Clean Air Act

This material does not contain any hazardous air pollutants.

This material does not contain any Class 1 Ozone depletors.

This material does not contain any Class 2 Ozone depletors.

Clean Water Act

None of the chemicals in this product are listed as Hazardous Substances under the CWA.

None of the chemicals in this product are listed as Priority Pollutants under the CWA.

None of the chemicals in this product are listed as Toxic Pollutants under the CWA.

OSHA

CAS# 7722-84-1 is considered highly hazardous by OSHA.

State

Hydrogen peroxide can be found on the following state right-to-know lists: California, New Jersey, Florida, Pennsylvania, Minnesota, Massachusetts.

Water is not present on state lists from CA, NJ, FL, PA, MN, or MA.

California No Significant Risk Level: None of the chemicals in this product are listed.

SECTION 16—Additional Information

MSDS Creation Date: 1/13/1995   Revision #9 Date: 9/30/1997

The information above is believed to be accurate and represents the best information currently available to us. However, we make no warranty of merchantability or any other warranty, express or implied, with respect to such information, and we assume no liability resulting from its use. Users should make their own investigations to determine the suitability of the information for their particular purposes. In no way shall Fisher be liable for any claims, losses, or damages of any third party or for lost profits or any special, indirect, incidental, consequential or exemplary damages, howsoever arising, even if Fisher has been advised of the possibility of such damages.

MATERIAL SAFETY DATA SHEET—Toluene

SECTION 1—Chemical Product and Company Identification

MSDS Name: Toluene
Synonyms: Methacide, methylbenzene, methylbenzol, phenylmethane, toluol
Company Identification: Fisher Scientific, 1 Reagent Lane, Fairlawn, NJ 07410
For information call: 201-796-7100
Emergency Number: 201-796-7100
For CHEMTREC assistance call: 800-424-9300
For International CHEMTREC assistance call: 703-527-3887

SECTION 2—Composition, Information on Ingredients

CAS#	Chemical Name	%	EINECS#
108-88-3	Benzene, methyl-	>99	203-625-9

Hazard Symbols: XN F
Risk Phrases: 11 20

SECTION 3—Hazards Identification

Emergency Overview

Appearance: colorless. Flash Point: 40°F

DANGER! Flammable liquid. May cause skin irritation. Harmful if inhaled. This substance has caused adverse reproductive and fetal effects in animals. May cause central nervous system depression. Aspiration hazard. May be absorbed through the skin.

POISON! May cause liver and kidney damage. Causes digestive and respiratory tract irritation. Harmful or fatal if swallowed. Causes eye irritation and possible transient injury.

Target Organs: Kidneys, central nervous system, liver

Potential Health Effects

Eye:	Causes eye irritation. May result in corneal injury. Vapors may cause eye irritation.
Skin:	May cause skin irritation. Prolonged and/or repeated contact may cause irritation and/or dermatitis. May be absorbed through the skin.
Ingestion:	Aspiration hazard. May cause irritation of the digestive tract. May cause effects similar to those for inhalation exposure. Aspiration of material into the lungs may cause chemical pneumonitis, which may be fatal.
Inhalation:	Inhalation of high concentrations may cause central nervous system effects characterized by headache, dizziness, unconsciousness, and coma. Inhalation of vapor may cause respiratory tract irritation. May cause liver and kidney damage. Vapors may cause dizziness or suffocation. Overexposure may cause dizziness, tremors, restlessness, rapid heart beat, increased blood pressure, hallucinations, acidosis, kidney failure.
Chronic:	Prolonged or repeated skin contact may cause dermatitis. May cause cardiac sensitization and severe heart abnormalities. May cause liver and kidney damage.

SECTION 4—First Aid Measures

Eyes:	Flush eyes with plenty of water for at least 15 minutes, occasionally lifting the upper and lower lids. Get medical aid immediately.
Skin:	Flush skin with plenty of soap and water for at least 15 minutes while removing contaminated clothing and shoes. Get medical aid if irritation develops or persists.
Ingestion:	Do NOT induce vomiting. If victim is conscious and alert, give 2–4 cupfuls of milk or water. Never give anything by mouth to an unconscious person. Possible aspiration hazard. Get medical aid immediately.
Inhalation:	Get medical aid immediately. Remove from exposure to fresh air immediately. If not breathing, give artificial respiration. If breathing is difficult, give oxygen.
Notes to Physician:	Causes cardiac sensitization to endogenous catecholamines which may lead to cardiac arrhythmias. Do NOT use adrenergic agents such as epinephrine or pseudoepinephrine.

SECTION 5—Firefighting Measures

General Information: Containers can build up pressure if exposed to heat and/or fire. As in any fire, wear a self-contained breathing apparatus in pressure-demand, MSHA/NIOSH approved (or equivalent), and full protective gear. Water runoff can cause environmental damage. Dike and collect water used to fight fire. Vapors may form an explosive mixture with air. Vapors can travel to a source of ignition and flash back. Flammable Liquid. Can release vapors that form explosive mixtures at temperatures above the flashpoint. Use water spray to keep fire-exposed containers cool. Water may be ineffective. Material is lighter than water and a fire may be spread by the use of water. Vapors may be heavier than air. They can spread along the ground and collect in low or confined areas. Containers may explode when heated.

Extinguishing Media: Use water spray to cool fire-exposed containers. Water may be ineffective. Do NOT use straight streams of water. For small fires, use dry chemical, carbon dioxide, water spray or regular foam. Cool containers with flooding quantities of water until well after fire is out. For large fires, use water spray, fog or regular foam.

Autoignition Temperature: 896°F (480°C)

Flash Point: 40°F (4.44°C)

NFPA Rating: health-2; flammability-3; reactivity-0

Explosion Limits: Lower: 1.1 Upper: 7.1

SECTION 6—Accidental Release Measures

General Information: Use proper personal protective equipment as indicated in Section 8.

Spills/Leaks: Avoid runoff into storm sewers and ditches which lead to waterways. Remove all sources of ignition. Absorb spill using an absorbent, noncombustible material such as earth, sand, or vermiculite. A vapor suppressing foam may be used to reduce vapors. Water spray may reduce vapor but may not prevent ignition in closed spaces.

SECTION 7—Handling and Storage

Handling: Wash thoroughly after handling. Use with adequate ventilation. Ground and bond containers when transferring material. Avoid contact with eyes, skin, and clothing. Empty containers retain product residue (liquid and/or vapor) and can be dangerous. Keep container tightly closed. Avoid contact with heat, sparks and flame. Avoid ingestion and inhalation. Do not pressurize, cut, weld, braze, solder, drill, grind, or expose empty containers to heat, sparks or open flames.

Storage: Keep away from heat, sparks, and flame. Keep away from sources of ignition. Store in a tightly closed container. Store in a cool, dry, well-ventilated area away from incompatible substances.

SECTION 8—Exposure Controls, Personal Protection

Engineering Controls: Use adequate general or local exhaust ventilation to keep airborne concentrations below the permissible exposure limits.

Exposure Limits	Chemical Name	ACGIH	NIOSH	OSHA - Final PELs
	Benzene, methyl-	50 ppm	100 ppm TWA	375 mg/m^3 TWA
		200 ppm TWA	188 mg/m^3	500 ppm IDLH
				C 300 ppm

OSHA Vacated PELs:

Benzene, methyl-: 100 ppm TWA; 375 mg/m^3 TWA

Personal Protective Equipment

Eyes: Wear appropriate protective eyeglasses or chemical safety goggles as described by OSHA's eye and face protection regulations in 29 CFR 1910.133.

Skin: Wear appropriate protective gloves to prevent skin exposure.

Clothing: Wear appropriate protective clothing to prevent skin exposure.

Respirators: Follow the OSHA respirator regulations found in 29 CFR 1910.134. Always use a NIOSH-approved respirator when necessary.

SECTION 9—Physical and Chemical Properties

Physical State:	Liquid
Appearance:	Colorless
Odor:	Sweetish odor—pleasant odor
pH:	Not available
Vapor Pressure:	10 mm Hg
Vapor Density:	3.1 (Air=1)
Evaporation Rate:	2.4 (Butyl acetate=1)
Viscosity:	0.59 cP at 68°F
Boiling Point:	232°F
Freezing/Melting Point:	-139°F
Decomposition Temperature:	Not available
Solubility:	0.6 mg/L H_2O at 68°F
Specific Gravity/Density:	0.9 (Water = 1)
Molecular Formula:	$C_6H_5CH_3$
Molecular Weight:	92.056

SECTION 10—Stability and Reactivity

Chemical Stability: Stable under normal temperatures and pressures

Conditions to Avoid: Incompatible materials, ignition sources, excess heat

Incompatibilities with Other Materials: Nitrogen tetroxide, nitric acid + sulfuric acid, silver perchlorate, strong oxidizers

Hazardous Decomposition Products: Carbon monoxide, carbon dioxide

Hazardous Polymerization: Has not been reported

SECTION 11—Toxicological Information

RTECS#:	CAS# 108-88-3: XS5250000
LD_{50}/LC_{50}:	CAS# 108-88-3: Inhalation, mouse: LC_{50} = 400 ppm/24H; Inhalation, rat: LC_{50} = 49 mg/m^3/4H; Oral, rat: LD_{50} = 636 mg/kg; Skin, rabbit: LD_{50} = 12,124 mg/kg
Carcinogenicity:	Benzene, methyl- —ACGIH: A4—Not Classifiable as a Human Carcinogen
IARC:	Group 3 carcinogen
Epidemiology:	No information available
Teratogenicity:	Specific developmental abnormalities included craniofacial effects involving the nose and tongue, musculoskeletal effects, urogenital and metabolic effects in studies on mice and rats by the inhalation and oral routes of exposure. Some evidence of fetotoxicity with reduced fetal weight and retarded skeletal development has been reported in mice and rats.
Reproductive Effects:	Effects on fertility such as abortion were reported in rabbits by inhalation. Paternal effects were noted in rats by inhalation. These effects involved the testes, sperm duct, and epididymis.
Neurotoxicity:	No information available
Mutagenicity:	No information available
Other Studies:	None

SECTION 12—Ecological Information

Ecotoxicity:	Bluegill LC_{50} =17 mg/L/24H Shrimp LC_{50} =4.3 ppm/96H Fathead minnow LC_{50}=36.2 mg/L/96H Sunfish (fresh water) TLm = 1,180 mg/L/96H
Environmental Fate:	From soil, substance evaporates and is microbially biodegraded. In water, substance volatilizes and biodegrades.
Physical/Chemical:	Photochemically produced hydroxyl radicals degrade substance.
Other:	None

SECTION 13—Disposal Considerations

Dispose of in a manner consistent with federal, state, and local regulations.

RCRA D-Series Maximum Concentration of Contaminants: None listed

RCRA D-Series Chronic Toxicity Reference Levels: None listed

RCRA F-Series: None listed

RCRA P-Series: None listed

RCRA U-Series: CAS# 108-88-3: waste number U220

CAS# 108-88-3 is banned from land disposal according to RCRA.

SECTION 14—Transport Information

US DOT	Shipping Name: TOLUENE
	Hazard Class: 3
	UN Number: UN1294
	Packing Group: II
IMO:	No information available
IATA:	No information available
RID/ADR:	No information available

Canadian TDG

Shipping Name:	TOLUENE
Hazard Class:	3(9.2)
UN Number:	UN1294
Other Information:	FLASHPOINT 4°C

SECTION 15—Regulatory Information

US FEDERAL

TSCA

CAS# 108-88-3 is listed on the TSCA inventory.

Health & Safety Reporting List: CAS# 108-88-3: Effective Date: October 4, 1982; Sunset Date: October 4

Chemical Test Rules: None of the chemicals in this product are under a Chemical Test Rule.

Section 12b: None of the chemicals are listed under TSCA Section 12b.

TSCA Significant New Use Rule: None of the chemicals in this material have a SNUR under TSCA.

SARA

Section 302 (RQ):	final RQ = 1,000 pounds (454 kg)
Section 302 (TPQ):	None of the chemicals in this product have a TPQ.
SARA Codes:	CAS # 108-88-3: acute, flammable
Section 313:	This material contains Benzene, methyl- (CAS# 108-88-3, >99%), which is subject to the reporting requirements of Section 313 of SARA Title III and 40 CFR Part 373.

Clean Air Act

CAS# 108-88-3 is listed as a hazardous air pollutant (HAP).

This material does not contain any Class 1 Ozone depletors.

This material does not contain any Class 2 Ozone depletors.

Clean Water Act

CAS# 108-88-3 is listed as a Hazardous Substance under the Clean Water Act (CWA).

CAS# 108-88-3 is listed as a Priority Pollutant under the CWA.

CAS# 108-88-3 is listed as a Toxic Pollutant under the CWA.

OSHA

None of the chemicals in this product are considered highly hazardous by OSHA.

STATE

Benzene, methyl- can be found on the following state right-to-know lists: California, New Jersey, Florida, Pennsylvania, Minnesota, Massachusetts.

WARNING: This product contains Benzene, methyl-, a chemical known to the state of California to cause birth defects or other reproductive harm.

California No Significant Risk Level: None of the chemicals in this product are listed.

SECTION 16—Additional Information

MSDS Creation Date: 1/04/1995   Revision #24 Date: 12/12/1997

The information above is believed to be accurate and represents the best information currently available to us. However, we make no warranty of merchantability or any other warranty, express or implied, with respect to such information, and we assume no liability resulting from its use. Users should make their own investigations to determine the suitability of the information for their particular purposes. In no way shall Fisher be liable for any claims, losses, or damages of any third party or for lost profits or any special, indirect, incidental, consequential or exemplary damages, howsoever arising, even if Fisher has been advised of the possibility of such damages.

Section 3C: Understanding Health Hazards
Instructor Notes

The objective of this section is for students to become familiar with the health hazard information provided on MSDSs. They review the MSDS health data for a number of solvents, select those posing the least health hazards, and explain their reasoning.

Student Pages

- Section 3C: Understanding Health Hazards—Background
- Section 3C: Understanding Health Hazards—Exercise

Possible Playout of the Section

Through a class discussion, students generate a list of substances they would term "toxic" and a list they would call "harmless." You may wish to set limits on the types of chemicals considered—e.g., common laboratory chemicals, solvents and reagents that they are familiar with, or common household products. Students read the Background, review and reevaluate their original lists of "toxic" and "harmless" substances, and discuss reasons for any changes they would make. Students then complete the Exercise.

Section 3C: Understanding Health Hazards

Background

Note: This section provides general recommendations for and information about MSDSs. However, you should always follow the procedures established by your organization's chemical hygiene plan.

Toxicology is the study of adverse effects of chemicals on living systems. Chemicals can have an immediate adverse effect (acute) or act over a long period of time (chronic). The basic premise of toxicology is that no substance is entirely safe and that all chemicals can cause harm to living systems if the amount encountered is great enough. Paracelsus (1493–1541) stated this concept nearly five centuries ago: "All substances are poisons; there is none which is not a poison. The right dose differentiates a poison...." For example, water, an essential ingredient for life, can be lethal if enough is ingested in a short time.

Defining "Poison," "Toxic," and "Toxin"

If all chemicals have some potential for harm, which ones do we label as toxic and which ones poisonous? The single most important factor is the relationship between the amount (or concentration) of the chemical encountered and the adverse health effect it produces. Each chemical has a range of concentrations that result in a graded effect between no effect and death. Toxicologists call this the dose-response relationship. Some chemicals, such as dioxin, will produce death in guinea pigs upon exposure to microgram doses. Another chemical may have no harmful effect even after a dose of many grams. Toxicologists have established the threshold dose for many chemicals—this is the concentration below which the chemical is not considered to be harmful.

The word "poison" is generally used to describe substances that cause human illness. In general use, a toxic substance is any material that is destructive or deadly. Health and safety professionals may use the term "toxic" more selectively than the term "poison" to describe chemicals for which certain levels of hazard have been demonstrated. The word "toxin" refers to a poisonous or toxic substance secreted by a living organism.

Health Risk Information Provided by MSDSs

MSDSs provide information about a variety of health hazards. These may include the following:

- acute (immediate) effects such as burns or unconsciousness;

- chronic (long-term) effects such as allergic sensitization, skin problems, or respiratory disease;

- whether the material is listed as a carcinogen by the Occupational Safety and Health Administration (OSHA), the International Agency for Research on Cancer Monographs (IARC), or the National Toxicology Program (NTP);

- limits to which a worker can be exposed (described in more detail below);

STUDENT PAGE

- primary routes of entry into the body;

- specific target organs likely to sustain damage; and

- medical problems that can be aggravated by exposure.

If the MSDS is for a mixture or solution, health hazard information must be provided for any hazardous ingredients that occur in parts of 1% or greater; for cancer-causing ingredients, the same information must be listed for occurrences in parts 0.1% or greater. However, it is important to note that the exposure limits and toxicological information that appear in the MSDS are for a 100% concentration (or as pure a state as possible) of each chemical listed. The individual chemicals in a mixture or solution are listed by Chemical Abstract Services (CAS) Registry Number.

The Exposure Controls section of an MSDS provides approved exposure limits based on inhalation of the chemical in the air. The threshold limit values (TLVs) established by the American Conference of Governmental Industrial Hygienists (ACGIH) refer to "airborne concentrations of substances and represent conditions under which it is believed that nearly all workers may be repeatedly exposed day after day without adverse health effects." (ACGIH, 1999) OSHA's name for a very similar value is permissible exposure limit (PEL). (For the purposes of this discussion, we will consider them to be interchangeable.) TLV is usually expressed as the number of parts of the compound in 1,000,000 parts of air (ppm) or the weight in milligrams in a cubic meter of air (mg/m^3). Two types of TLV measures are used. TLV-TWA (threshold limit value-time weighted average) is the "time-weighted average concentration for a conventional 8-hour workday and a 40-hour workweek, to which it is believed that nearly all workers may be repeatedly exposed, day after day, without adverse effect." (ACGIH, 1999) TLV-STEL (threshold limit value-short term exposure limit) is the highest safe concentration for a 15-minute exposure during on eight-hour workday. In an MSDS, TLV with no other designation (such as TWA or STEL) usually means TLV-TWA. In general, substances with TLVs of less than 50 ppm should be handled in a fume hood.

The Health Hazard Data section of an MSDS includes toxicity measures, which are indications of how harmful the chemical can be. Toxicity is indicated as the mean dose or concentration causing death in 50% of laboratory animals tested. Typically, small mammals such as mice, rats, and rabbits are used as test animals. It is believed that these mammals' physiological responses to chemicals are similar to those of humans. A measurement called LC_{50} is the concentration of a hazardous material in air that is expected to kill 50% of a group of test animals when given as a single respiratory exposure in a specific time period. LD_{50} is a single dose, other than inhalation, that causes death in 50% of an animal population from exposure to a hazardous substance. LD_{50} may be based on oral or skin contact doses. LD_{50} and LC_{50} are expressed in units that are difficult for many people to relate to: milligrams or grams per kilogram of body weight. The values are easier to understand if converted to the probable lethal human dose for an average-sized person, say, 150 pounds. However, the relationship between LC_{50} and LD_{50} data on test animals and the effect of the test chemical on humans is not fully known.

Toxicology data based on animal studies should be used only as a guide to rank the toxicity of various chemicals.

Probable Lethal Oral Dose for Humans			
Animal LD_{50}	Ingested by 150-pound adult	Toxicity classification	Example
less than 5 mg/kg	less than 7 drops	extremely toxic	mercuric chloride LD_{50} oral (rat): 1 mg/kg
5–50 mg/kg	7 drops to 1 teaspoonful	highly toxic	potassium cyanide LD_{50} oral (rat): 10 mg/kg arsenic trioxide LD_{50} oral (rat): 15.1 mg/kg
50–500 mg/kg	1 teaspoonful to 1 ounce	moderately toxic	formaldehyde LD_{50} oral (rat): 100 mg/kg
500 mg/kg–5 g/kg	1 ounce to 1 pint	slightly toxic	aspirin (acetylsalicylic acid) LD_{50} oral (rat): 1,000 mg/kg table salt (NaCl) LD_{50} oral (rat): 3,000 mg/kg
Over 5 g/kg	more than 1 pint	practically nontoxic	glycerin LD_{50} oral (rat): 12,600 mg/kg
Adapted with permission from *Prudent Practices in the Laboratory*. Toxicological data from *Flinn Scientific Chemical & Biological Catalog Reference Manual 1999*, pp 910–911.			

Carcinogenic Toxicity

An MSDS also includes information on the cancer-causing potential of chemicals. Although cancer ranks as the second most common cause of death in the United States, the process of carcinogenesis is not clearly understood. As a result, several problems arise in evaluating the carcinogenic potential of chemicals. Human health is affected by a wide range of factors, including occupation, genetic disposition, and lifestyle. Therefore, the relationship between any one exposure and the onset of cancer is difficult to determine. Second, many cancers are latent responses—that is, the disease may not show itself until years after the initial exposure. Despite these difficulties, several organizations have systems for classifying chemicals according to carcinogenic potential. You may see one or more of these classifications on the MSDS.

The IARC, a segment of the World Health Organization, classifies carcinogens into three categories. Group 1 carcinogens are chemicals for which sufficient epidemiological evidence exists that they cause cancer in humans. Asbestos, benzene, and vinyl chloride fall into this category. Group 1 materials are sometimes called "known human carcinogens." The second category is divided into two sub-groups. Materials in Group 2 are commonly called "probable or possible carcinogens." Group 2A materials (probable carcinogens) are those shown to cause cancer-like changes in experimental animals or mammalian cells. This category includes polychlorinated biphenyls (PCBs), formaldehyde (gas), and silica dust. Group 2B materials (possible carcinogens) show cancer-causing effects in animal studies or genetic mutations in specific strains of bacteria. Group 3 includes a wide variety of chemicals for which there is no direct evidence of carcinogenesis in humans, although epidemiological correlations indicate that they may cause cancer. Many common chemicals, including hydrogen peroxide and toluene, are included in this category.

The EPA, OSHA, and ACGIH have also established coding systems for classifying the carcinogenicity of various chemicals. ACGIH uses the following designations:

- A1—Confirmed Human Carcinogen—carcinogenic to humans based on strong epidemiological or clinical evidence

- A2—Suspected Human Carcinogen—carcinogenic to experimental animals at dose levels (or other criteria) considered relevant to worker exposure

- A3—Animal Carcinogen—carcinogenic in experimental animals at relatively high dose levels (or other criteria) not considered relevant to worker exposure

- A4—Not Classifiable as a Carcinogen—inadequate data on which to classify

- A5—Not Suspected as a Human Carcinogen—not suspected as carcinogenic to humans based on properly conducted epidemiological studies in humans

Under OSHA, substances may be classified as select carcinogens or potential carcinogens. Select carcinogens (see Section 6E: Working with Hazardous Substances) include substances listed under IARC Group 1, or listed in the category "known to be carcinogens" in the Annual Report on Carcinogens published by the NTP. Some IARC Group 2 substances can also be select carcinogens if they cause statistically significant tumor incidence in experimental animals in accordance with exposure criteria established by OSHA. Exposure criteria for known or select carcinogens are usually reported with an Effective Dose value measured in milligram of substance per kilogram of body weight per unit time.

Some chemicals also have mutagenic effects, meaning that they can cause genetic changes. These chemicals are called mutagens. Mutagens may or may not also be carcinogens. Teratogens are chemicals known to cause birth defects. Teratogens differ from mutagens in that they disrupt normal fetal development but may not cause genetic changes. For example, ethyl alcohol, which is known to cause fetal alcohol syndrome, is not shown to have mutagenic effects. This information is also reported on the MSDS. New information about the carcinogenic, mutagenic, and teratogenic effects of chemicals is continually being published. Staying up to date with this information is extremely important.

Section 3C: Understanding Health Hazards
Exercise

This Exercise will familiarize you with the health hazard information provided on MSDSs, particularly terms related to exposure limits, and help you interpret exposure limits for a list of common solvents. Many laboratory operations involve the selection of an appropriate solvent. Glassware cleaning, azeotropic distillations, liquid-liquid extractions, and chromatography separations (thin layer, column, HPLC) all require choosing the "best" solvent. Traditionally, "best" has more often referred to the cost and the operational effectiveness, with little or no consideration given to the related exposure and potential health issues. However, today health and safety considerations have taken on additional importance. For example, citrus oils are replacing halogenated solvents in many cleaner applications.

The table "MSDS Health Hazard Data for Some Common Solvents" on the following page provides hazard data for different categories of solvents. Review the table, making sure you understand the meanings of PEL-TWA, STEL, LD_{50}, and LC_{50}. For each compound category listed in the table, identify the solvent that poses the least health risk. Explain your choices.

A=Toluene; because it has the highest PEL-TWA, STEL, and oral LD_{50}.

HC=Pentane; because it has the highest PEL-TWA and STEL of the hydrocarbon solvents. No LD_{50} indicated.

OH=Ethyl alcohol, because it has the highest PEL-TWA and oral LD_{50}.

−X=Probably methylene chloride; because it has the highest PEL-TWA on the list.

OT=Acetone; because it has the highest PEL-TWA and STEL on the list.

MSDS Health Hazard Data for Some Common Solvents						
	Compound Category	OSHA Limits		LD_{50} (mg/kg) (lethal dose)	LC_{50} (ppm) (lethal concentration)	FLP ($°F$) (flash point under normal conditions)
		PEL-TWA (ppm) (permissible exposure limit)	STEL (ppm) (short-term exposure limit)			
Acetone	OT	750	1,000	oral (rat): 5,800 skin (rabbit): 20,000	inhale (rat): 50,100	0
Acetic Acid	OT	10	15 (ACGIH)	oral (rat): 3,530 skin (rabbit): 1,060	inhale (mouse): 5,620	104
Acetonitrile	OT	40	60	oral (rat): 2,730 skin (rabbit): 1,250	inhale (rat): 7,551	42
Acrylonitrile	OT	2	N/A	oral (rat): 78 skin (rabbit): 250	N/A	32
Benzene	A & HC	1	5	oral (rat): 3,306	inhale (rat): 10,000	12
Butyl Alcohol	OH	100	N/A	N/A	N/A	84
Carbon Disulfide	OT	4	12	oral (rat): 3,188 skin (rabbit): 2,550	inhale (rat): 25,000	-22
Carbon Tetrachloride	–X	2	N/A	oral (rat): 2,350 skin (rabbit): 5,070	inhale (rat): 8,000	non-flammable
Chlorobenzene	A & –X	75	N/A	oral (rat): 2,290	N/A	75
Chloroform	–X	2	N/A	oral (rat): 908	inhale (rat): 48	non-flammable
Dimethylformamide	OT	10	N/A	oral (rat): 2,800 skin (rabbit): 4,720	inhale (mouse): 9,400	153
Ethyl Acetate	OT	400	N/A	oral (rat): 5,620	inhale (rat): 1,600	24
Ethyl Alcohol	OH	1,000	N/A	oral (rat): 7,060	inhale (rat): 20,000	55
Diethyl Ether	OT	400	500	oral (rat): 1,215 skin (rabbit): 250	inhale (rat): 73,000	-49
Formalin (Formaldehyde 37% solution)	OT	1	2	oral (rat): 800 skin (rabbit): 270	inhale (rat): 590	185
n-Hexane	HC	50	N/A	oral (rat): 28,700	N/A	-9.4
Isopropyl Alcohol	OH	400	500	oral (rat): 5,045 skin (rabbit): 12,800	N/A	53
Methyl Alcohol	OH	200	250	oral (rat): 5,628 skin (rabbit): 15,840	inhale (rat): 64,000	52
Methylene Chloride	–X	500	N/A	oral (rat): 1,600	inhale (rat): 88,000	non-flammable
Pentane	HC	600	750	N/A	N/A	-57
Tetrahydrofuran	OT	200	250	oral (rat): 2,816	inhale (rat): 21,000	6
Toluene	A & HC	200	500	oral (rat): 5,000 skin (rabbit): 12,124	N/A	40
Xylene	A & HC	100	150	oral (rat): 4,300	inhale (rat): 5,000	78

Key for Compound Categories: A = aromatic solvent, HC = hydrocarbon solvent, OH = alcohol, –X = Halogenated hydrocarbon solvent, OT = other solvent. Toxicity data taken from the National Institute of Health's National Toxicology Program Chemical Health and Safety Database. http://ntp-server.niehs.nih.gov/ (accessed March 20, 2000).

Section 3D: Finding MSDSs on the Web
Instructor Notes

The objective of this section is for students to use Web-browsing software to locate MSDSs on the Internet and apply the information from the MSDSs to answer questions about certain chemicals.

Student Pages

- Section 3D: Finding MSDSs on the Web—Background
- Section 3D: Finding MSDSs on the Web—Exercise

Possible Playout of the Section

The students will need to know how to use the Web-browsing software on the available computers. Students are asked to find the MSDS(s) for the chemical(s) you assign them. We recommend you select from the 88 chemicals that have Laboratory Chemical Safety Summaries (LCSSs) in *Prudent Practices in the Laboratory* (see list below) so that they can compare the MSDS to the LCSS in Exercise 3E. These LCSSs are also available at *www.qrc.com/hhmi/science/labsafe/lcss/start.htm*.

acetaldehyde	acetic acid	acetone
acetonitrile	acetylene	acrolein
acrylamide	acrylonitrile	aluminum trichloride
ammonia (anhydrous)	ammonium hydroxide	aniline
arsine	benzene	boron trifluoride
bromine	*tert*-butyl hydroperoxide	butyl lithium
carbon disulfide	carbon monoxide	carbon tetrachloride
chlorine	chloroform	chloromethyl methyl ether
chromium trioxide	cyanogen bromide	diazomethane
diborane	dichloromethane	diethyl ether
diethylnitrosamine	dimethyl sulfate	dimethyl sulfoxide
dimethylformamide	dioxane	ethanol
ethyl acetate	ethylene dibromide	ethylene oxide
fluorides (inorganic)	fluorine	formaldehyde
hexamethylphosphoramide	hexane	hydrazine
hydrobromic acid and hydrogen bromide	hydrochloric acid and hydrogen chloride	hydrogen
hydrogen cyanide	hydrogen fluoride and hydrofluoric acid	hydrogen peroxide
hydrogen sulfide	iodine	lead and its inorganic compounds
lithium aluminum hydride	mercury	methanol
methyl ethyl ketone	methyl iodide	nickel carbonyl
nitric acid	nitrogen dioxide	osmium tetroxide
oxygen	ozone	palladium on carbon
perchloric acid	phenol	phosgene
phosphorus	potassium	sodium and potassium hydride
pyridine	silver and its compounds	sodium
sodium azide	sodium cyanide and potassium cyanide	sodium and potassium hydroxide
sulfur dioxide	sulfuric acid	tetrahydrofuran
toluene	toluene diisocyanate	trifluoroacetic acid
trimethylaluminum	trimethyltin chloride	

After the Exercise, students might share with the class where they found their MSDSs and whether they would recommend the website to others. You may wish to have students compile a class list of the most useful Universal Resource Locators (URLs) for locating MSDSs.

Section 3D: Finding MSDSs on the Web
Background

Note: This section provides general recommendations for and information about MSDSs. However, you should always follow the procedures established by your organization's chemical hygiene plan.

Paper copies of MSDSs are available from chemical suppliers. However, today nearly any MSDS can be downloaded from the Web. The information about the Web included here is intended to provide a brief introduction to Internet resources for MSDSs. Throughout, we presume that you have a web browser and access to information on how to use it. The following is a brief review of how to navigate through the Web to find MSDSs.

Using a Search Engine

The Internet contains a vast amount of information scattered around the world. How can you find what you are looking for? One answer is search engines. Search engines are websites that enable you to enter specific subjects or key words to locate information on the Web. Your web browser may provide access to one or more search engines, often through a button marked "Net Search" or something similar. You can also reach search engines by entering the URL directly. For entering URLs directly, your browser will provide a menu choice such as "Open Location." The URLs for a few popular search engines are listed below.

Excite	*www.excite.com/*	AltaVista	*altavista.com/*
Yahoo!	*www.yahoo.com/*	WebCrawler	*webcrawler.com/*
Infoseek	*infoseek.go.com/*		

Once you have accessed the search engine, type in the word(s) or subject you are looking for and click on the indicated button to begin the search. A page will return to you with the results of your search. To view a site, click on the underlined text in the listing. Your web browser will display the page. Many times your search will return thousands of references, presented in smaller groups and often listed in order of relevance to your request. However, you may not get what you expect. After a search, try a few of your listings to see if you are getting what you intended. If not, revise your search words and try again. Many search engines contain hints on refining your search criteria. Also, since different search engines may be linked to different websites, a new search with a different search engine may yield different results.

Internet Resources for MSDS Information

You can find MSDS sites on your own using search engines as described above, but we have provided a few URLs to get you started. Please note that although these URLs were correct when this book was published, the locations of webpages can change. Use a search engine as needed to update the provided URLs.

URLs for MSDS Information		
Site	URL*	Description
Vermont SIRI	http://hazard.com	Provides over 180,000 online searchable MSDSs plus other safety links and information.
Cornell University	http://www.pdc.cornell.edu	Contains about 325,000 online searchable MSDSs. Choose Health and Safety. From Health and Safety page, choose Material Safety Data Sheets.
Fisher Scientific	http://fishersci.com	Includes MSDSs and order information. Choose the Fisher Chemical or Acros Organic catalog. Search to find the chemical you want, then click on the MSDS logo.
University of Kentucky	http://www.ilpi.com/msds/index.html	Includes general MSDS information plus links to many other sites offering MSDSs.
*All were accessed March 1, 2000.		

Searching Tips

When navigating from site to site via hyperlinks, you can easily lose track of your location. If you need to return to a site where you found great information, you may not be able to find it easily again! To avoid this problem, keep track of your sources as described below. You may also want to print out a copy of the homepage for your files. If possible, when you print the homepage, specify to include the page's URL, date, and any other important information on the printout as well. You also may want to use the bookmark feature on your browser.

Keep in mind that the Internet is an unmoderated, unreviewed information source. You must evaluate the reliability and quality of the information you locate. Often, the identity of the sponsoring organization can help you judge the value of the information. In evaluating information, you must also consider the purpose for the website. For example, commercial sites may present information with a slant toward their sales agenda. Lobbying organizations also have their own particular agendas. When you have found what you believe to be a reliable information source, write down the following information:

- the name of the search engine employed and the keywords used to find the information,
- the URL,
- the author of the webpage (if given),
- the organization that is represented, and
- a brief description of the information on the site.

If you will be using the same computer for web searches at a later date, you may also wish to store links to some websites in your web browser. These stored links are called bookmarks or favorites. Your web browser probably offers a menu choice for bookmarking. Typically, you select a command such as "add bookmark" when viewing a webpage to which you plan to return. The link to that page is stored in a list of bookmarks that you can access later.

Section 3D: Finding MSDSs on the Web
Exercise

In this Exercise, you will try several of the URLs provided in the Background and view one or more MSDSs at each site. You may want to bookmark these in your browser for easy reference. Use a search engine to locate additional URLs that provide MSDSs and make a note of these URLs or bookmark them. The objective of the Exercise is for you to become comfortable with retrieving an MSDS for any given chemical from the Internet. Once your instructor assigns your chemical(s) to you, locate and print two different MSDSs for your assigned chemical(s) from the Web. Make sure that one MSDS is in the 16-part ANSI format and one is the older (nonstandard) type.

Review the MSDSs for safety considerations and record the values for the following: flash point, PEL-TWA, STEL, specific gravity, vapor pressure, lower explosive limit, upper explosive limit, NFPA ratings (health, flammability, reactivity), and CAS Registry Numbers and UN numbers. Briefly explain why knowing each of these values is important. If you do not understand the meaning of any of these values, look them up in a chemical reference.

Answer the following questions based on the information contained in the MSDSs.
Answers will vary based on the assigned chemical.

1. Is your assigned chemical more dense than water? Give a reason for your answer.

2. Are vapors from your assigned chemical less dense than air? Give a reason for your answer.

3. If someone were using a Bunsen burner in the laboratory, would you need to take special precautions when using your assigned chemical? Explain.

4. What personal protective equipment would you need for working with your assigned chemical? Explain.

5. In the event of a small spill of your assigned chemical in the laboratory, what steps should you follow to clean it up?

6. When using your assigned chemical in the laboratory, what should you do to ensure a safe working environment for yourself and your colleagues?

7. If you accidentally ingest some of your assigned chemical, what should you do?

Section 3E: Evaluating MSDSs
Instructor Notes

To become familiar with critiquing MSDSs, students will learn to assess MSDSs for completeness using a set of evaluation questions.

Student Pages

- Section 3E: Evaluating MSDSs—Background
- Section 3E: Evaluating MSDSs—Exercise

Possible Playout of the Section

Students read the Background and complete the Exercise. You may wish to have the class discuss what factors might affect the quality, reliability, and usefulness of the information found on MSDSs or discuss how laboratory workers can best evaluate the information found in MSDSs and apply this information to their own situation.

Section 3E: Evaluating MSDSs

Background

Note: This section provides general recommendations for and information about MSDSs. However, you should always follow the procedures established by your organization's chemical hygiene plan.

Although valuable, the usefulness of MSDSs in a laboratory setting is limited in the following ways:

- OSHA does not require the recommended ANSI format for MSDSs; therefore, MSDSs provided by different manufacturers may have different formats. Although the same required information should be on all MSDSs, quickly finding something specific can be difficult if you do not know exactly where to look. This can be a problem in an emergency.

- The voluntary ANSI standard provides format consistency, but it does not ensure the quality and degree of completeness of the information on the MSDS.

- MSDSs may contain vague language and broad generalizations.

- MSDSs may contain internal inconsistencies.

- MSDSs were originally designed for health and safety professionals, such as industrial hygienists, firefighters, and other emergency response personnel, not laboratory workers. Also, they are written to address the largest scale in which the material could be used, such as manufacturing operations. Thus, the hazard warnings and handling procedures on MSDSs often are not appropriate in the context of laboratory-scale work.

- MSDSs are written for the form and concentration of the material as manufactured and sold. This may be significantly different from the form and concentration of the material as used in the laboratory.

- MSDSs typically list every possible health hazard associated with a material without indicating which health hazards are most significant or most likely to occur. Laboratory workers can have difficulty distinguishing highly hazardous materials from relatively harmless ones.

- MSDSs can be for single substances, mixtures, or solutions. An MSDS provides exposure limits and toxicological data for 100% concentrations of all the hazardous chemicals it covers, even if the MSDS is for a mixture or solution.

Because of the limitations of MSDSs, the National Research Council has developed a chemical safety information format specifically to meet the needs of laboratory workers. These documents, called Laboratory Chemical Safety Summaries (LCSS), do not replace MSDSs; rather, they are intended to supplement them.

Section 3E: Evaluating MSDSs

Exercise

In this Exercise, you will critically review the clarity and completeness of the MSDSs you retrieved from the Internet and answer the following questions based on information from the Background. Based on your answers, decide whether your MSDSs are of good or poor quality.

Answers will vary based on the assigned chemical.

1. Does your MSDS have blank fields?

2. Does the MSDS list an emergency phone number, issue date, and responsible party?

3. Are symptoms of overexposure described in complex or confusing medical terms?

4. Do you have incomplete or missing TLV/PEL values?

5. Is specific personal protective equipment information missing?

6. Does the MSDS contain obvious inconsistencies?

7. Does the MSDS report that significant information is unknown?

8. If the MSDS provided to you in a work setting was of poor quality, what would you do?

Your instructor will provide you with an LCSS for your assigned chemical. Compare the LCSS with the two MSDSs and answer the following questions:

9. Which is easier to use: the MSDS or the LCSS? Why?

10. What information does an MSDS have that an LCSS does not? Is this information important to a laboratory worker? Why or why not?

11. Does the LCSS provide any information that was missing on the MSDSs? What?

12. Does the LCSS clarify any of the information provided by the MSDSs? How?

Chapter 4
Using Protective Equipment

The tragic mercury poisoning death of Professor Karen E. Wetterhahn of Dartmouth College on June 8, 1997 has lessons for chemists who think that disposable latex gloves protect against everything. While working with dimethylmercury in a fume hood, Dr. Wetterhahn spilled one or several drops on her disposable latex gloves during a transfer procedure. Three months later she began to experience nausea and then more serious symptoms. About six months after the accident she fell into a coma and subsequently died. [See "Section 4A: Laboratory Incident Reports" for more on this story.]

CHEM 13 News, September 1997

Chapter 4 Objectives

Pertinent Voluntary Industry Standards

- *VIS L2 Standards*—Choose and demonstrate the use of protective equipment to be used in a variety of situations (e.g., eye wear, special clothing).

- *VIS (L2.04):01*—Select and demonstrate the use of appropriate personal protective equipment (PPE) for a variety of situations involving hazardous chemicals, including but not limited to: corrosive, explosive, biological, and volatile.

Objectives of the Chapter

Students will learn

- what type of PPE is available,

- when and where to use PPE, and

- the limitations of PPE.

Why Your Students Need to Understand Protective Equipment

In academic laboratories you may see students avoiding goggles and other personal equipment because they say the glasses are hot or look stupid, or other protective equipment is uncomfortable. However, when they enter the workplace laboratory, students will find that avoiding PPE because it is inconvenient, uncomfortable, or unflattering is not tolerated. A manager of safety and training for a chemical manufacturer states that "Most of the young chemists and lab techs we get here straight out of college or who work here as summer interns have to have extensive safety training and external motivation to understand . . . the importance of PPE." Concern for employee health, avoidance of the costs associated with injuries, and the need to comply with OSHA regulations drive industry regulations regarding PPE.

In order to comply with OSHA, industrial laboratories should have detailed standard operating procedures that outline required use of PPE. However, some laboratories may not have these procedures. Your students need to understand that they have a personal responsibility for protecting their health by knowing how to select and use personal protective equipment to lower their exposure to hazards.

Overview of the Chapter

Chapter 4: Using Protective Equipment contains five sections. General background information and complete instructions for the exercises and labs are provided. The Background pages are intended only to provide basic background information on personal protective equipment for your students. We recommend that you extend this background information by having students find and read the pertinent sections of the books in the safety reference library (described in Chapter 1). You may wish to have students briefly report on the information they find during the monthly safety meetings.

Section 4A: Getting to Know Protective Equipment

This section introduces students to the importance of protective equipment and the regulations governing their use in the laboratory and workplace. To initiate interest in the topic, students read and discuss Laboratory Incident Reports concerning accidents that resulted from not using or misusing protective equipment. Students should be aware that the chapter does not cover all types of protective equipment. Many laboratories have other protective equipment specialized to their operations.

Section 4B: Eye and Face Protection

This section increases student awareness of the differences among protective devices for the eye and face and to motivate students to make informed decisions when choosing the appropriate protective devices.

Section 4C: Skin Protection

The objective of this section is to help students understand the importance of skin protection in the laboratory and learn how to select appropriate laboratory gloves. In the "Glove Degradation" and "Glove Permeation" Labs, students evaluate changes in glove materials when they are exposed to test chemicals.

Section 4D: Clothing Selection

Students become familiar with appropriate clothing selections for laboratory work and determine the chemical resistance of several blends of fabric to various chemicals.

Section 4E: Fume Hood Safety

This section helps students to understand the types of fume hoods, when to use them, and how to operate them safely.

References for Chapter 4

"Acid-in-the-Eye Simulation"; *ChemFax* #801; Flinn Scientific: Batavia, IL, 1996.

Arrotti, G. "The Eye Protection Program." *Occupational Health and Safety*, Aug 1995, 28–31.

Arrotti, G. "Fit the Gloves." *Occupational Health and Safety*, May 1995, 50–54.

Blayney, M.; Nierenberg, D.W.; Winn, J.S. "Handling Dimethylmercury." *Chemical & Engineering News,* May 12, 1997, *75* (19), 7.

Breazeale, W.H.; Ramsey, H. "Contact Lenses." *Chemical & Engineering News,* June 1, 1998, *76* (22), 6.

"Choosing Protective Eyewear"; EZ FACTS™ Safety Info by Fax; Lab Safety Supply: Janesville WI, 1998. (This is available free online at www.labsafety.com.)

Eagle Insurance Group Safety Meeting Outline, www.eig.com/smodex.htm#PPE; (accessed March 22, 2000).

Eye and Face Protection Standard. *Code of Federal Regulations,* 29 CFR 1910.133, 1999. (This is available free online at www.osha.gov.)

"Eye-Damage Simulation"; *ChemFax* #10062; Flinn Scientific: Batavia, IL, 1992.

FabricLink Website. Fabric University. Characteristics of Textile Fibers. www.fabriclink.com/ Characteristics.html (accessed March 6, 2000).

Flinn Fax!; 95 (4), Flinn Scientific: Batavia, IL, 1995.

Gorman, C.; *Working Safely with Chemicals in the Laboratory,* 2nd ed.; Genium: Schenectady, NY, 1997.

Hand Protection Standard. *Code of Federal Regulations,* 29 CFR 1910.138, 1999. (This is available free online at www.osha.gov.)

Hofstader, R.; Chapman, K. *Foundations for Excellence in the Chemical Process Industries: Voluntary Industry Standards for Chemical Process Industries Technical Workers;* American Chemical Society: Washington, DC, 1997.

Improving Safety in the Chemical Laboratory: A Practical Guide, 2nd ed.; Young, J.A., Ed.; John Wiley & Sons: New York, 1991; Chapter 18.

Lab Safety Supply, Inc. website. Tech Info. EZ Facts. Personal Protection. Chemical Protective Gloves. Document Number 191. www.labsafety.com/commerce/techhelpctr/thezfacts/ezf191.htm (accessed Feb 24, 2000).

Lab Safety Supply, Inc. website. Tech Info. EZ Facts. Personal Protection. Choosing Protective Eyeware. Document Number 125. www.labsafety.com/commerce/techhelpctr/thezfacts/ezf125.htm (accessed Feb 24, 2000).

"Latex Did Not Prevent Poisoning." *CHEM 13 NEWS,* Sept 1997, 8.

Learning By Accident; Mojtabai, F.; Kaufman, J., Eds.; Laboratory Safety Institute: Natick, MA, 1997; p 26, 56.

Long, J. "Mercury Poisoning Fatal to Chemist." *Chemical & Engineering News,* June 16, 1997, *75* (24), 11–12.

Mikell, W.; Fuller, F. "Good Practices for Safe Hood Operation." *Journal of Chemical Education,* Feb 1988, *65* (2).

Nagel, M.C. "A Latex Glove Alert." *Chemical Health and Safety,* 1997, *4* (6), 14–18.

Nakagawara, V.B. "Functional Model of an Eye Protection Program." *Professional Safety,* Nov 1989, 24–27.

NIOSH Pocket Guide to Chemical Hazards, June 1997; NIOSH: Cincinnati, OH.

Prudent Practices in the Laboratory; National Academy: Washington, DC, 1995.

Respiratory Protection Standard. *Code of Federal Regulations,* 29 CFR 1910.134, 1999. (This is available free online at www.osha.gov.)

Rozelle, W.N. "*Textile World* Manmade Fiber Chart 1998." Intertec: Atlanta, GA, 1998.

Safety in Academic Chemistry Laboratories; American Chemical Society: Washington, DC, 1995, 11–12; Appendix II.

Segal, E.B. "Contact Lenses and Chemicals." *Chemical Health and Safety,* 1997, *4* (3), 33–37.

"Skin Safety," *Healthlines: Special Report.* Spence, 1991.

Stull, J.O. "Risk Assessment for Protective Clothing." *Occupational Hazards,* Jan 1996, 47–48.

Stull, J.O. "Selecting Chemical Protective Clothing." *Occupational Health and Safety,* Dec 1995, 20–22.

Wortham, S.D. "Learn the ABCs of PPE." *Safety + Health,* Aug 1995, 48–51.

Section 4A: Getting to Know Protective Equipment
Instructor Notes

This section introduces students to the importance of protective equipment and the regulations governing their use in the laboratory and workplace. To initiate interest in the topic, students read and discuss Laboratory Incident Reports concerning accidents that resulted from not using or misusing protective equipment. Students should be aware that the chapter does not cover all types of protective equipment. Many laboratories have other protective equipment specialized to their operations.

Student Pages

- Section 4A: Getting to Know Protective Equipment—Laboratory Incident Reports
- Section 4A: Getting to Know Protective Equipment—Background

Possible Playout of the Section

Students read the Laboratory Incident Reports and Background and discuss the following questions:

1. What are the similarities and differences between the incidents described in the Laboratory Incident Reports?

2. What conclusions can you draw about safety from these incidents?

3. How might each of these incidents have been prevented?

Laboratory Incident Reports

Incident Description: Gloves

The tragic mercury poisoning death of Professor Karen E. Wetterhahn of Dartmouth College on June 8, 1997, has lessons for chemists who think that disposable latex gloves protect against everything. While working with dimethylmercury in a fume hood, Dr. Wetterhahn spilled one or several drops (estimated to total 0.1 to 0.5 mL) on her disposable latex gloves during a transfer procedure. Three months later she began to experience nausea and then more serious symptoms. About six months after the accident she lapsed into a coma and subsequently died.

Subsequent tests showed that dimethylmercury penetrates disposable latex gloves in 15 seconds or less, and perhaps instantaneously. A severely toxic dose of 100 to 200 mg of dimethylmercury requires absorption of less than 0.1 mL of liquid. Upon hearing of this case, a chemist at the University of Waterloo added his experience that dichloromethane—a common organic solvent—penetrates latex gloves within seconds. He can feel the slight stinging sensation of the spilled solvent even though no hole appears in the latex. He suspects that many liquids, though probably not aqueous solutions, pass easily through latex gloves.

CHEM 13 News, September 1997, and *Chemical and Engineering News,* May 12 and June 16, 1997

Incident Description: Goggles

While cleaning equipment for checking out of the lab using dichromate cleaning solution, an analytical chemistry student dropped the bottle of cleaning solution in the sink, splashing the liquid in his eyes. He was wearing contact lenses and no goggles. A lab assistant walked in on the scene and started to flush the student's eyes. Another student got the instructor who asked about safety goggles and contact lenses. The lab assistant and instructor removed the lenses and kept flushing. The student was then taken to the hospital where his eyes were treated and bandaged. The student was angry at himself, realizing he knew better and was at fault for the accident. Fortunately, because of the immediate attention he received, he lost no sight and had no scars.

Learning by Accident, page 56

Incident Description: Fume Hoods

I was working as a chemical technologist for a company that sold radioactive isotopes to researchers around the world. Part of my job was to dispense liquid solutions of radioactive iodine (I-125). The lab contained several hoods with sophisticated filter systems. We wore protective eye glasses, gloves, and lab coats. Despite the precautions taken, the worker in the hood next to me accidentally spilled some I-125 on the apron of the hood. He had the door of the hood open too far and the suction was poor. I-125 is very volatile, and he did not tell me what he had done so I could evacuate the lab. We both suffered from overexposure as we inhaled the radioactive fumes. The I-125 ended up in my thyroid gland. I discovered something was wrong when, upon leaving the lab, I checked myself with a G.M. tube, and it screamed when it got to my throat.

Learning by Accident, page 26

Section 4A: Getting to Know Protective Equipment
Background

This section provides general recommendations for and information about protective equipment. However, you should always follow the protective equipment policy established by your organization's chemical hygiene plan.

In academic laboratories, you may see your fellow students avoiding goggles, gloves, and other personal protective equipment because they say they are hot, look stupid, make it hard to handle or read equipment, or are otherwise inconvenient. You may have done this yourself. However, when you enter the laboratory workplace, avoiding protective equipment because it is inconvenient, unflattering, or uncomfortable, will not be tolerated and can lead to dismissal. Concern for employee health, avoidance of the costs associated with injuries, and the need to comply with OSHA regulations result in strict industry rules regarding protective equipment. At many industrial facilities, safety glasses are required when you enter the facility.

You will find that protective equipment plays a critical role in hazard control. The principles of hazard control call for risk reduction by reducing the probability of an accident and reducing the severity of accidents when they happen. A level of zero risk is unattainable, but risk management reduces exposure to hazards by following a hierarchy that outlines hazard controls in descending order of desirability. A hazard control hierarchy typically includes 1) administrative controls (such as training of employees), 2) engineering controls (such as exhaust ventilation systems), and 3) controls by personal protective equipment (such as goggles and gloves). *Safety + Health* (1995) states, "Safety professionals know that engineering and administrative controls are preferable to personal protective equipment (PPE) to keep workers safe. But hazards are an inherent part of some jobs and production techniques, and PPE offers a way to enable workers to more safely coexist with them."

Chemical and Physical Hazards

Protective equipment is intended to provide a physical barrier to chemical exposure via most routes of entry, as well as a barrier to physical hazards such as explosions and heat. These types of hazards and the role of protective equipment in minimizing them are discussed in the following paragraphs.

How Chemicals Can Enter the Body

Chemicals can enter and damage your body via four primary routes: absorption through skin or eyes, injection, inhalation, and ingestion. When properly used, PPE can protect against entry of chemicals by inhalation or absorption and provide some protection against accidental injection. Protection from ingestion requires your proactive attention.

Absorption of chemicals through the skin or eyes can cause anything from a slight irritation to permanent damage. Some toxins that enter the body this way (such as phenol) can enter your bloodstream and cause damage to your nerves, liver, kidneys, or red blood cells. Gloves and aprons are used to provide protection from absorption. *You will learn about correct goggle use in Section 4B: Eye and Face Protection. Section 4C: Skin Protection explores correct glove use.*

Injection of a chemical can occur accidentally when handling sharp items such as hypodermic needles or broken glass. Gloves and aprons, depending on what they are made of, provide some protection against injection. Proper laboratory clothing (long pants and sleeves, shoes without openings, etc.) also provides a measure of protection from injection. *You will become familiar with appropriate clothing for laboratory work in Section 4D: Clothing Selection.*

The most likely route for chemical exposure is inhalation. Gases, vapors, dust, and fumes in your close proximity may enter your lungs when you breathe. Some chemicals, such as hydrogen fluoride, ammonia, chlorine, and asbestos, damage your lungs directly. Others, such as hydrogen cyanide, enter your bloodstream and affect your body's ability to take up oxygen. Alcohols and solvents can affect the central nervous system. If operating and used properly, fume hoods usually provide adequate protection from inhalation of fumes, dust, and vapors. However, in some instances, specially fitted respirators may be required. *You will explore the correct use of fume hoods in Section 4E: Fume Hood Safety.*

Ingestion of chemicals may occur by eating, drinking, smoking with contaminated hands or in a contaminated area, or otherwise placing materials in your mouth. This is why laboratories have strict rules against eating, drinking, and applying cosmetics in the laboratory or storing food for consumption in the laboratory refrigerator.

Physical Hazards to the Body

Your body is vulnerable to many hazards besides entry of chemicals into the body. Physical hazards in the laboratory include flying airborne particles, such as glass from an exploding flask; molten metal and other very hot materials; very cold materials; acids and other corrosives that can burn skin; injurious light; radiation; heavy falling or rolling objects, such as gas cylinders; electric shock; and noise. To minimize the risk from these hazards, you must know when and how to use protective equipment such as safety shields, gloves, aprons, hearing protectors, radiation suits, and steel-toed shoes.

Your Rights and Responsibilities

You should know your rights and responsibilities in the workplace regarding PPE. In order to comply with OSHA, academic and industrial laboratories should have detailed standard operating procedures that outline required PPE. However, they may not have these procedures. You need to assume personal responsibility for protecting your health by knowing how to select and use PPE to lower your exposure to hazards. You should know why PPE is required; what equipment to use, when to use it and how to use it; limitations of the equipment; and how to care for, maintain, and repair it.

A college chemistry instructor posted the following statement to an e-mail list: "Some of our students arrive with the idea that a strict door-to-door eye protection policy is kid stuff: in the real world individuals choose if and when to wear PPE." An industry safety officer replied, "Those people are the ones who will NOT last in a good company."

Section 4B: Eye and Face Protection
Instructor Notes

The objective of this section is to increase student awareness of the differences among protective devices for the eye and face and to motivate students to make informed decisions when choosing the appropriate protective devices.

Student Pages

- Section 4B: Eye and Face Protection—Background
- Section 4B: Eye and Face Protection—Exercise

Possible Playout of the Section

Provide students with copies of your institution's regulations for eye and face protection including the chemical hygiene plan and/or standard operating procedures for handling chemicals. Have students read these and the Background and discuss the implications this information has for student work and for full-time laboratory employees. You may also wish to discuss the changing ideas about contact lens use in the laboratory.

Two instructor-led demonstrations ("Acid-in-the-Eye Simulation" and "Eye-Damage Simulation," provided on the following pages) dramatically illustrate the points made in this section. Students are typically shocked to learn that even a very short exposure time can lead to significant eye damage. Students then complete the Exercise, in which they choose the types of eye and face protection to use with several different chemicals.

Safety for the Demonstrations

The demonstrations in this book do not routinely include special safety precautions for the chemicals or equipment that might be used. As the course instructor, it is expected that you will both follow SOPs and local regulations for conducting any demonstration and discuss these precautions with your class to reinforce the building of a safety culture.

Setup Notes for the Demonstrations

Acid-in-the-Eye Simulation

Materials
- marker
- Petri dish
- raw eggs or egg whites
- overhead projector
- Beral pipets (one for each solution)

- 6 M HCl (aq)
- (optional) 6 M NaOH (aq)
- appropriate personal protective equipment (PPE)

Eye-Damage Simulation

Materials
- 2 (or more) cow eyes, fresh (or other fresh mammalian eyes)
- 250-mL beakers, one for each cow eye
- wax marking pencil
- concentrated sulfuric acid (18 M H_2SO_4)
- (optional) pipet for dispensing H_2SO_4 (conc)
- stopwatch
- sink with running water
- appropriate personal protective equipment (PPE)

Presentation of the Demonstrations

Acid-in-the-Eye Simulation

Put on appropriate PPE and reinforce to students the reasons for using the PPE.

Place raw egg white in the Petri dish. Place the Petri dish on the overhead projector stage for the class to observe. (To emphasize the point of the demonstration, you may want to draw an eyeball on the overhead and place the Petri dish of egg white over it.)

Using the pipet, place several drops of 6 M HCl on the raw egg white. It will immediately become opaque. (Note, acid solutions of less than 6 N also react with the egg white, but the reaction is less dramatic.) Try to "undo" the damage by gently rinsing the egg white with water. (The egg white remains denatured.)

You may wish to repeat the demonstration with a 6 M NaOH solution to show the effect of a concentrated base. (Concentrated solutions of sodium hydroxide also denature the egg whites, turning them white.)

Discuss the fact that egg whites and human eyes both contain proteins that denature when subjected to concentrated solutions of strong acids. The destruction of the three-dimensional structure (denaturing) of the protein changes its properties and is frequently irreversible.

After the demonstration, the egg whites should be cautiously rinsed with water and then disposed of according to state and local ordinances.

Eye-Damage Simulation

Put on the appropriate PPE required for handing concentrated H_2SO_4 and discuss reasons for using the PPE, including requirements of SOP or MSDS.

Rinse the cow eyes, if necessary, and place each of them in a separate 250-mL beaker. Using the wax pencil, label the beakers "control" and "30 seconds." Pass around to students the cow-eye beaker labeled "control." Tell students to observe the color and clarity of the eye.

Starting with the beaker labeled "30 seconds," cautiously pour about 1 mL H_2SO_4 (conc) onto the cow eye, taking extreme care to avoid any contact with the H_2SO_4 (conc). Leave the acid on for about 30 seconds and then cautiously (to prevent splattering) pour cold water over the cow eye. Place in a sink, and thoroughly rinse the eye with cold running water for about 5 minutes, draining the water and flushing the eye with water several more times. Then, place the eye into an appropriate container to pass around for students to observe.

You may wish to repeat the treatment with different cow eyes, allowing them to be exposed to the H_2SO_4 (conc) for less time. Be sure to rinse as before and allow students to observe. Our tests showed that cow eyes treated for 30 seconds and 15 seconds showed severe damage. There was substantial damage with a 5-second exposure, and a "rinse immediately" test showed noticeable damage, but not as severe as longer exposures.

Thoroughly rinse all cow eyes and dispose of them according to local ordinances. Do not place the cow eyes in a garbage can that students may have access to.

Acknowledgments

Acid-in-the-Eye Simulation adapted from *ChemFax* #801; Flinn Scientific: Batavia, IL, 1996.

Eye-Damage Simulation adapted from *ChemFax* #10062; Flinn Scientific: Batavia, IL, 1992.

Background

This section provides general recommendations for and information about protective equipment. However, you should always follow the protective equipment policy established by your organization's chemical hygiene plan.

Laboratory eye and face protection is designed to protect you from chemical splashes and projectiles. According to the Eagle Insurance Group, 1,000 eye injuries occur in American workplaces every day, representing over $3,000,000 annually in medical expenses, lost production, and workers compensation costs. A survey by the Bureau of Labor Statistics showed that 60% of eye injuries happen because the worker was not wearing any eye protection. The other 40% of injured workers were wearing improper eye protection. In injuries involving flying objects, nearly 75% of the objects causing eye injury are no more than 0.5 mm in diameter. Serious damage to your eyes could completely change your life, as illustrated by one safety slogan: "You can eat with false teeth and you can dance with a wooden leg, but you can't see with a glass eye."

OSHA Eye Protection Standards

Under OSHA's Eye and Face Protection Standard (29 CFR 1910.133), employers must provide suitable eye and face protection for employees exposed to flying projectiles, dusts and mists, molten metals or other chemical splashes or splatters, or injurious radiation. Employers must provide appropriate eye protection that is reasonably comfortable under the conditions of use, that fits securely without interfering with the wearer's vision or movements, is durable, and is kept clean and in good repair. ANSI provides detailed standards for each type of protective eyewear.

Types of Eye and Face Protection

The type of eye protection you wear must fit the hazards you may encounter on the job. These hazards include both flying objects of various sizes and chemical splashes. Nearly every industrial laboratory requires some type of protective eyewear at all times. Safety glasses with side shields are considered by some to be the bare minimum protection. Regular prescription glasses do not take the place of safety glasses and do not offer adequate protection. However, prescription safety glasses are available, and more flattering styles are available than in the past. True prescription safety glasses have the lens maker's mark on the lenses of the glasses and on the frame.

Safety goggles provide a much higher level of protection than safety glasses. They protect your eyes from projectiles and, if designed to do so, will also protect against chemical splashes. The best type of goggles for protection from projectiles and chemical splashes are the indirect ventilated goggles. Direct ventilated goggles should never be worn in the chemistry laboratory. Safety goggles should be used if chemical splashes could occur or when working with glassware under reduced or increased pressure, compressed gases, potentially explosive compounds, or glassware under high temperatures. Different eye protection is required to limit exposure to potentially injurious light sources, ultraviolet radiation, lasers, and temperature extremes. Workers who are required to wear eye protection for long periods of time may also want to consider comfort features, such as facial contours, lens tinting, ventilation, and fog-free lenses. Safety eyewear should be washed frequently with mild soap and water to provide continuous good visibility for the user. The following table explains the features of different protective eyewear.

Explanation of Protective Eyewear Equipment	
	Safety Spectacles or Glasses Safety spectacles have protective side shields to shield your eyes from minor impacts.
	Direct Ventilated Goggles Goggles with direct vents offer protection from impact only. They fit snugly around the eye area to prevent flying objects from striking your eyes. Direct-ventilated goggles offer more comfort because they allow air to flow in and out to reduce the chance of fogging.
	Indirect Ventilated Goggles Goggles with indirect vents allow air to move freely in and out without allowing splash particles in. They offer the same impact protection as the direct-ventilated goggles. Because there isn't as much space for the air to move in and out, lens fogging may be a slight problem, and you may want to consider an anti-fog lens coating to alleviate any potential problems.
	Non-ventilated Goggles Non-ventilated goggles have no holes for air to seep through. They offer a higher level of protection against vapors and fumes and can be used to keep harmful vapors out of sensitive eyes. These goggles must have an anti-fog coating to keep them from steaming up while you work.
	Face Shields A face shield is a protective device worn to shield the wearer's face from hazards. Face shields are secondary protectors only and must be worn with safety goggles, as stated in ANSI Z87.1–1989 and OSHA.

Face and Neck Protection

Face shields are another type of protective equipment. They are available in two basic styles: covering the face only and covering the face and neck. Face shields provide more protection to the face and neck than goggles alone and are recommended when working with highly corrosive liquids, with glassware under reduced or increased pressure, with glassware at high temperatures, and when an explosion or implosion is possible. However, never consider a face shield a primary source of protection. Safety goggles must be worn with the face shield, especially in the chemistry laboratory. The following table shows the eyewear appropriate for different tasks.

Face shields must always be used with goggles, as face shields alone will not provide adequate protection.

Eyewear Selection Chart		
Operation	Hazards	Protective Devices (see key)
Impact Chipping, grinding, machining, masonry work, riveting, and sanding	Flying fragments, large chips, particles, sand, dirt, machine parts, etc.	1, 2, 3, 4, 5
Heat Furnace operations, pouring, casting, hot dipping, gas cutting, and welding	Hot sparks	1, 2, 3, 4, 5
	Splash from molten metals	4, 5 (for severe exposure)
	High-temperature exposure	5 (plus appropriate goggles)
Chemical Acids and chemical handling, degreasing, and plating	Splash	3, 4, 5 (for severe exposure)
	Irritating mists	4
Dust Wordworking, buffing, and general dusty conditions	Nuisance dust	3, 4, 5
Key	1: Spectacle, 2: Direct ventilated goggles, 3: Indirect ventilated goggles, 4: Non-ventilated goggles, 5: Face shields	
Adapted with permission from *EZ Facts Safety Information, Document #125*		

Safety Shields

Safety shields are small barriers constructed out of laminated safety glass or transparent polymeric material placed between the reaction and the person manipulating the reaction. Safety shields offer additional protection against chemical splashes and flying materials. Safety shields do not replace protective eyewear; they should be used in addition to face shields and goggles. Safety shields protect only against minor blasts or explosions from small amounts of reactants. To use safety shields correctly, make sure shields are weighted or mechanically secured to prevent tipping, and surround the reaction with the shields so that those near the reaction are protected as well as the person manipulating the apparatus. Remember, someone may be working on the other side of the lab bench.

The Debate Over Contact Lenses in the Laboratory

Until recently, the prevailing opinion concerning contact lenses has been to recommend (and sometimes require) that contact lenses not be permitted in the laboratory. However, many assumptions about the risks of contact lenses in the laboratory are being challenged with new evidence. Some common beliefs are that contact lenses can hold foreign materials against the cornea, can absorb and retain chemical vapors, and can make damage from corrosive chemical splashes more severe. Given this list of presumed hazards, policies prohibiting the use of contact lenses in the laboratory are understandable.

The actual risks or disadvantages to wearing contact lenses in the laboratory include those listed here. However, most of these problems will not arise if proper eye protection is worn.

- Dust or dirt may get trapped under hard contact lenses.

- Some solvents can interact with the materials of both hard and soft contact lenses.

- Contact lenses may cloud or discolor when exposed to certain chemicals.

- Some gases (such as ammonia) and vapors (such as isopropyl alcohol and ethyl alcohol vapors) can be absorbed by the lenses and irritate or damage the eyes.

- Some contact lenses prevent oxygen from reaching the eyes.

- Workers who wear contact lenses are at a greater risk when working alone because if irritation occurs, it is necessary to remove the contact lens from the eye and this becomes more difficult when no one is around to help. However, workers should not be alone when working with hazardous materials anyway.

However, contact lenses are now generally thought to create no additional hazards for the wearer who uses appropriate eye protection. In some situations, they may even be advantageous. Wearing contact lenses in the laboratory provides the following benefits:

- Contact lenses can provide better vision than eyeglasses for some people.

- Contact lenses may be more comfortable than eyeglasses.

- Contact lenses provide better peripheral vision than eyeglasses.

- Soft UV-absorbing lenses protect the cornea from optical radiation.

- Contact lenses are better for work that involves optical instruments.

- Studies of various gases, vapors, fumes, and smoke found in laboratory settings have shown that contact lenses do not absorb chemicals like a sponge but instead may actually help protect the eye from potentially dangerous airborne substances.

- Examples of exposure to corrosive chemicals by contact-lens wearers point to no additional risks. In some cases, they support evidence that contact lenses protect against severe eye damage.

Based on recent study and review of the issue, the American Chemical Society "is of the consensus that contact lenses can be worn in most work environments, provided the same approved eye protection is worn as required of other workers in the area." (*Chemical & Engineering News,* June 1998) If you wear contact lenses, follow the standard operating procedures set for your laboratory regarding this issue. Some academic institutions and workplaces have clearly defined rules about the use of contact lenses; if yours is one, be sure to follow them. If you are left to make your own decision or are involved in making a decision about this issue, make sure you take into consideration the benefits as well as the actual (not perceived) risks so that you can make a decision that you and your laboratory can be comfortable with.

Section 4B: Eye and Face Protection
Exercise

In this Exercise, you will evaluate the eye and face protection needed when working with various chemicals. The objective of the Exercise is to help you apply what you have learned about selecting appropriate PPE for the eyes and face.

Consider each of the chemicals listed in the table below and the stated hazards. Name the type of eye and/or face protection you would choose and explain your choice. Note that local and state regulations may be more restrictive than OSHA standards. Always consult your employer's standard operating procedures and follow local and state regulations. The toxicity information presented here (excerpted from Laboratory Chemical Safety Summaries in *Prudent Practices in the Laboratory*) is only for the purpose of evaluating eyewear choices. Before using any of these chemicals, read the complete MSDS and/or LCSS and follow locally established protocols for their use.

Toxicity of Selected Chemicals	
Substance	Selected Toxicity
Acetonitrile CAS 75-05-8	Acetonitrile is slightly toxic by acute exposure through ingestion, skin contact, and inhalation and can be converted by the body to cyanide. It can be severely irritating to the eyes and slightly irritating to the skin.
Ammonia CAS 7664-41-7	Ammonia gas is extremely corrosive and irritating to the skin, eyes, nose, and respiratory tract. Eye contact with ammonia vapor is severely irritating. Eye exposure to ammonia vapor, mists, and liquid can cause severe damage, which may result in permanent eye injury and blindness.
Dichloromethane (methylene chloride) CAS 75-09-2	Dichloromethane is only slightly toxic if exposure occurs orally or by inhalation but contact of the compound with the eyes causes painful irritation and can lead to conjuctivitis and corneal injury if not promptly removed by washing.
Diethyl Ether CAS 60-29-7	Ether's toxicity level is low. It is mildly irritating to the eyes and skin but does not generally cause irreversible damage.
Hydrogen CAS 1333-74-0	Hydrogen is practically non-toxic. Exposure to high concentrations of this gas may cause loss of consciousness due to lack of oxygen.
Iodine CAS 7553-56-2	Iodine can be highly toxic if inhaled. Iodine vapor causes severe eye irritation after exposure to relatively high concentrations for more than 2 minutes.
Perchloric acid CAS 7601-90-3	Perchloric acid is a highly corrosive substance with moderate acute toxicity. It is a severe irritant to the eyes, mucous membranes, and upper respiratory tract.
Pyridine CAS 110-86-1	Pyridine has a low acute toxicity. It can irritate the eyes and skin. It is readily absorbed, leading to possible systemic effects such as nausea, vomiting, and central nervous system damage.
Toluene CAS 108-88-3	Toluene has a low acute toxicity but may cause eye, skin and respiratory tract irritation. Contact of liquid toluene with the eyes causes temporary irritation.

Section 4C: Skin Protection
Instructor Notes

The objective of this section is to help students understand the importance of skin protection in the laboratory and learn how to select appropriate laboratory gloves. In the "Glove Degradation" and "Glove Permeation" Labs, students evaluate changes in glove materials when they are exposed to test chemicals.

Student Pages

- Section 4C: Skin Protection—Background
- Section 4C: Skin Protection—Lab: Glove Degradation
- Section 4C: Skin Protection—Lab: Glove Permeation

Possible Playout of the Section

After reading the Background, students discuss the variety and recommended usage of glove materials for different laboratory applications.

In the "Glove Degradation" lab, students evaluate changes in mass when glove materials are exposed to test chemicals. Because of the extreme hazards associated with handling concentrated H_2SO_4, we recommend that the instructor do these tests as a demonstration to prevent student contact with this chemical. You may also want to specifically discuss the changes in the nature of the glove materials. For example, H_2SO_4 (conc) can cause the glove to char, discolor, curl up, or become brittle.

In the "Glove Permeation" lab, students use a pH meter to measure changes in the pH inside finger assemblies that are immersed in the test chemicals. Note that the Procedure calls for the glove samples to be scrubbed and rinsed before use. This removes any powder, which could otherwise affect the pH and skew the results. In the discussion of the student results, we suggest you caution students about reading too much into changes in pH that are less than 0.5, which is at the upper limit of the experimental error for the activity.

For both labs we recommend that you divide the class into groups to test as many glove types and chemical test reagents as you have available. The class can pool and compare their results.

Emphasize to students that the procedures they are using are not as rigorous as industry tests and that their results may vary widely from the recommendations provided by glove manufacturers. Also emphasize that students should not use their own data as a basis for selecting appropriate laboratory gloves. Rather, they should always use the recommendations of the glove manufacturer. These labs are intended only to give a general idea of the types of tests used to evaluate gloves and to familiarize students with the characteristics of some types of gloves.

Safety

The labs in this book do not routinely include special safety precautions for the chemicals or equipment that might be used. As the course instructor, it is expected that you will provide students with access to SOPs, MSDSs, and other resources they need to safely work in the laboratory while meeting all regulatory requirements. Before doing any of the labs from this book or from other sources, we recommend that you regularly review special handling issues with students, allow time for questions, and then assess student understanding of these issues.

Remind the students that no glove, no matter how highly rated, is a substitute for proper chemical handling and procedures. Also remind them always to avoid contact with reagents, even while wearing appropriate safety equipment, and to follow the SOPs established by your institution.

Setup Notes for Glove Degradation

Materials per student
- appropriate personal protective equipment (PPE)

Materials per group
- forceps
- glass dropper or disposable glass pipet (if the assigned chemical is not in a dropper bottle)
- beakers for each glove type to be tested
  - 2, approximately 250 mL (1 for the 5-minute test, 1 for the 30-minute test)
  - 1, approximately 600 mL or larger
- 1 of the glove types listed
  - nitrile
  - latex
  - neoprene
  - vinyl
- masking tape and pen for labels
- stopwatch
- paper towels

Materials per class
- balance
- test chemicals, such as
  - 6 M H_2SO_4
  - acetone
  - hexane
  - concentrated sulfuric acid (18 M H_2SO_4) for instructor demo only

Setup Notes for Glove Permeation

Materials per student
- appropriate personal protective equipment (PPE)
- 3 glove fingers, cut from one of the following glove types
  - nitrile
  - latex
  - neoprene
  - vinyl
- lab brush
- 3, 100-mL beakers (150- or 250-mL beakers can be substituted if needed)
- 10-mL graduated cylinder
- tape and marker for labels
- test chemicals, such as
  - 6 M HCl
  - 1 M HCl
  - 6 M NaOH
  - 1 M NaOH
- distilled water
- 3 PVC tubes, 1-inch diameter, about 3½ inches long
- 3 clamps
- 1 ring stand
- pH meter
- stopwatch
- large beaker

Representative Student Data: Glove Degradation

Note that these data are from student tests, and that the results may vary widely from manufacturer data. Also, the results are likely to vary with the brand of glove tested.

		Nitrile		Latex		Neoprene		Vinyl	
		5 min	30 min	5 min	30 min	5 min	30 min	5 min	30 min
18 M H_2SO_4	Initial Mass (g)	1.187	1.256	1.456	1.332	0.973	1.003	1.002	1.023
	Final Mass (g)	1.545	1.597	1.407	2.657	0.966	1.035	1.157	1.025
	Percent Mass Change	30.2%	27.2%	3.4%	99.5%	0.72%	3.2%	15.5%	0.20%
	Rating	Fair	Fair	Excellent	Not Rated	Excellent	Excellent	Good	Excellent
6 M H_2SO_4	Initial Mass (g)	1.134	1.505	1.296	1.309	1.130	1.422	1.430	1.396
	Final Mass (g)	1.112	1.500	1.238	1.245	1.126	1.412	1.429	1.396
	Percent Mass Change	1.9%	0.3%	4.5%	4.9%	0.4%	0.7%	0.07%	0%
	Rating	Excellent	Excellent	Excellent	Excellent	Excellent	Excellent	Excellent	Excellent

Table title: Sample Glove Degradation Data

acetone	Initial Mass (g)	1.328	1.395	1.468	1.285	1.042	1.200	1.013	1.288
	Final Mass (g)	1.265	1.291	1.426	1.223	1.032	1.181	0.622	0.735
	Percent Mass Change	4.7%	7.5%	2.9%	4.8%	0.96%	1.6%	38.6%	42.9%
	Rating	Excellent	Excellent	Excellent	Excellent	Excellent	Excellent	Poor	Poor
hexane	Initial Mass (g)	1.024	1.166	1.331	1.483	1.236	1.244	1.240	1.144
	Final Mass (g)	1.020	1.156	1.309	1.400	1.224	1.220	0.835	0.747
	Percent Mass Change	0.4%	0.86%	1.7%	5.6%	0.97%	1.9%	32.7%	34.7%
	Rating	Excellent	Excellent	Excellent	Excellent	Excellent	Excellent	Poor	Poor

Representative Student Data: Glove Permeation

These data are from student tests; results may vary widely from manufacturer data. Also, the results are likely to vary with the brand and thickness of glove tested.

Sample Glove Permeation Data					
		Initial pH	5 min pH	30 min pH	60 min pH
distilled H$_2$O (control)	Latex	7.2	7.1	7.1	7.1
	Nitrile	7.7	7.6	7.6	7.6
	Neoprene	7.1	7.1	7.1	7.1
	Vinyl	7.1	7.0	7.0	7.0
1 M HCl	Latex	7.1	7.0	7.0	6.8
	Nitrile	7.9	7.8	7.6	7.6
	Neoprene	7.1	7.1	7.1	7.1
	Vinyl	7.0	7.0	7.0	7.0
6 M HCl	Latex	7.1	6.9	6.8	6.6
	Nitrile	7.6	7.4	7.4	7.3
	Neoprene	7.1	7.0	7.0	6.9
	Vinyl	7.0	7.0	6.9	6.8
1 M NaOH	Latex	7.1	8.6	9.0	9.0
	Nitrile	6.8	7.9	9.2	9.6
	Neoprene	7.2	8.2	8.7	9.2
	Vinyl	7.3	8.3	8.6	8.8
6 M NaOH	Latex	7.0	8.2	9.3	9.3
	Nitrile	7.0	8.2	9.3	9.6
	Neoprene	6.7	8.1	9.0	9.0
	Vinyl	6.9	7.9	9.0	9.4

Section 4C: Skin Protection

Background

This section provides general recommendations for and information about protective equipment. However, you should always follow the protective equipment policy established by your organization's chemical hygiene plan.

If asked to name the largest organ of your body, would your answer be "skin"? You may never have thought about skin as an organ, but it's a very important one—and quite large. If spread out flat, the skin of an average-sized person would cover an area of about 3 feet by 7 feet. Skin helps regulate body temperature, prevents harmful bacteria from entering the body, excretes salts and liquids, keeps the body from drying out, and enables you to experience the sense of touch. Your skin also absorbs ultraviolet rays from the sun, enabling your body to produce vitamin D.

Skin has three layers: the epidermis, dermis, and hypodermis. The outer layer of the skin, the epidermis, is about as thick as the piece of paper on which this page was printed. The top layer of the epidermis is composed mostly of dead cells that provide the first line of defense against germs and harmful chemicals. Underneath these dead cells lies a layer of rapidly dividing cells that harden and die as they migrate to the outer layer. The middle layer of the skin, the dermis, is 15 to 40 times as thick as the epidermis. The dermis contains blood vessels, oil and sweat glands, hair follicles, and nerves. Below the dermis lies the hypodermis, which is made up of fat cells that provide insulation and act as shock absorbers.

OSHA Skin Protection Standards

Under OSHA's Hand Protection Standard (29 CFR 1910.138), employees must wear appropriate hand protection when exposed to hazards such as skin absorption of harmful substances, severe cuts, lacerations or abrasions, punctures, chemical or thermal burns, and harmful temperature extremes. Because no single material is suited for all applications, hand protection materials must be tested for each application or task. Selection of the appropriate material must be based on a performance evaluation measuring factors such as degradation rate, breakthrough time, and permeation rate. This evaluation is considered relative to the tasks to be performed, conditions present, duration of use, and hazards and potential hazards identified.

Causes of Skin Damage

In the chemistry laboratory, skin is vulnerable to damage from extremes of heat or cold, sharp objects, and corrosive or irritating chemicals. Exposed skin also provides a route of entry for toxic chemicals. Skin diseases are one of the leading job-related illnesses in the United States today. Over 1 million workers every year suffer from some type of skin problem acquired on the job. Minor skin injury damages only the epidermis. Epidermal damage may allow hazardous chemicals to go deeper and cause more damage. Major skin injury involves all three layers of the skin.

Some of the more common types of chemicals that damage skin tissue are listed below:

- The soaps and detergents you use daily remove dirt from your skin, but they may also remove some of your skin's protective layers. Exposure to harsh cleaning agents may irritate your skin and make it more vulnerable to other damage.

- Solvents such as acetone and toluene are very effective in dissolving grease. However, they can also remove your skin's oil and moisture, leaving your skin red and blistered. Also, many solvents can enter the body by penetrating through intact skin.

- Corrosive chemicals, such as strong acids and bases, can burn, eat away, and destroy skin tissue.

- Some metal compounds can cause serious allergic reactions and red ulcerated skin. Some people are highly allergic to these compounds and may have a long-term allergic reaction to them.

- Some chemicals such as phenol, hydrogen fluoride, and dimethyl sulfoxide enter the body directly through the skin and have chronic effects on other body systems.

Laboratory Gloves

Because you use your hands so much while working in a laboratory, your hands are constantly at some risk. Gloves provide essential protection from an injury or illness due to chemical contact with skin. *Prudent Practices in the Laboratory* states that "Proper protective gloves must be worn when handling hazardous chemicals, toxic materials, materials of unknown toxicity, corrosive materials, rough or sharp-edged objects, and very hot or very cold objects." Proper protective gloves are not always typical rubber latex gloves. In fact, many times natural rubber latex gloves do not provide adequate protection from skin hazards in the lab. (Latex may also be problematic because of latex allergies. See "Allergic Reaction to Latex Gloves" later in this section.) The glove material must be appropriate for the type of chemicals being used and the tasks being performed. Gloves must also fit properly and be free of defects that you may not be able to see just by looking at them. You can check for pinholes or defects in many gloves by inflating them with air from the laboratory air line and then submerging the inflated glove in water and looking for bubbles.

Choosing the Appropriate Laboratory Glove

Wearing the wrong gloves for the job may be worse than wearing no gloves at all because if a chemical penetrates a glove, the glove material may end up prolonging contact between the chemical and your skin, increasing the injury or damage. This is not to say that it is appropriate to go without gloves; rather, it emphasizes the need to know what materials you are working with and how to protect yourself. One glove might offer the highest level of chemical resistance but little or no cut resistance; another might offer the highest cut resistance with little or no chemical resistance. Cloth gloves offer protection from low temperatures but are inappropriate for handling cryogenic fluids because the liquid permeates the fabric and can cause severe frostbite. Often a combination of gloves is necessary to provide adequate protection. So how do you choose the right glove or combination of gloves for the tasks you need to accomplish? Most glove manufacturers offer standard test data on their glove materials, but looking at these data may not provide enough information because data usually only exist for pure chemicals. Combinations of chemicals require testing of glove material before a choice of glove is made.

Two values are commonly used to rate glove suitability for various chemicals: degradation and permeation. Degradation refers to deterioration of the glove material. Permeation is the ability of a chemical to move through the glove, even if no damage to the material occurs. For example, "because latex is readily permeated by carbon disulfide, a hand covered by a latex glove immersed in carbon disulfide would receive constant wetting by this toxic chemical, which would then be absorbed through the skin." *(Prudent Practices in the Laboratory)* Several sites on the Internet offer extensive information on glove types and the protection they offer against hundreds of chemicals. Two are *www.bestglove.com/* and *www.inform.umd.edu/CampusInfo/Departments/EnvirSafety/ls/gloves.html.* The table below presents some glove types and an overview of the protection from chemical exposure they offer.

Chemical Protective Gloves	
Glove	Applications
Butyl®	A synthetic rubber material that offers the highest permeation resistance to gas and water vapors. Especially suited for use with esters and ketones.
Neoprene	A synthetic rubber material that provides excellent tensile strength and heat resistance. Neoprene is compatible with some acids and caustics. It has moderate abrasion resistance.
Nitrile	A synthetic rubber material that offers chemical and abrasion resistance— a very good general-duty glove. Nitrile also provides protection from oils, greases, petroleum products, and some acids and caustics.
PVC (polyvinyl chloride)	A synthetic thermoplastic polymer that provides excellent resistance to most acids, fats, and petroleum hydrocarbons. Good abrasion resistance.
PVA™ (polyvinyl alcohol)	A water-soluble synthetic material that is highly impermeable to gases. Excellent chemical resistance to aromatic and chlorinated solvents. This glove cannot be used in water or water-based solutions.
Viton®	A fluoroelastomer material that provides exceptional chemical resistance to chlorinated and aromatic solvents. Viton is very flexible but provides minimal resistance to cuts and abrasions.
Silver Shield®/4H™	A lightweight, flexible, laminated material that resists permeation from a wide range of toxic and hazardous chemicals but provides virtually no cut resistance. Usually worn under or over other types of heavier gloves.

In addition to providing protection from chemical exposure, some types of gloves offer other protection. For example, cloth gloves protect against dirt, abrasions, wood slivers, and low temperatures; metal mesh protects against cuts, rough materials, and sharp-edged tools; and rubber acts as an electrical insulator.

Allergic Reaction to Latex Gloves

Repeated exposure to natural rubber latex gloves may cause the wearer to develop an allergic reaction to latex. People with a history of other allergies are at a greater risk, especially those who have allergies to avocados, potatoes, bananas, tomatoes, chestnuts, kiwi fruit, and papaya. The most common symptoms are skin rashes or irritation; tearing, redness, and itching of the eyes; sneezing; and tightness of the throat. Some symptoms can be more severe such as shortness of breath, chest pain, palpitations, fast pulse, or anaphylaxis. Persons with known latex allergies must avoid contact. If laboratory gloves are needed and latex becomes the choice of the laboratory, be aware that latex does pose a risk of allergic reaction, especially if worn for long periods of time. People who become sensitized to latex can trigger a reaction by contact with or inhalation of other latex products in or outside of the laboratory. Such products include toy balloons, rubber bands, erasers, automobile tires, and rubber boots. The National Institute of Occupational Safety and Health (NIOSH) website has an informative article titled "Preventing Allergic Reactions to Natural Rubber Latex in the Workplace" *(www.cdc.gov/niosh/latexalt.html)*.

Care of Laboratory Gloves

Be sure to remember to decontaminate or wash your gloves appropriately before removing them. If they cannot be cleaned, dispose of them according to procedures at your institution. Always remove your gloves before handling objects like doors, keyboards, pens, and phones, and leave your gloves in a work area where they will not touch uncontaminated objects. Don't forget to wash your hands after taking off your gloves. When disposing of gloves, turn them inside out when taking them off. This will help contain any materials remaining on the gloves.

Section 4C: Skin Protection
Lab: Glove Degradation

In this Lab, you submerge samples of glove material in a test chemical, calculate the percent change in mass, and rate the glove type according to that change in mass. This procedure is adapted from tests conducted by glove manufacturers to assign ratings to their products.

Safety

In a laboratory setting, you are ultimately responsible for your own safety and for the safety of those around you. In order to more closely resemble an actual workplace, specific safety information is not routinely provided to you for the labs in this book. As discussed throughout the book, many resources are available to assist you in obtaining the information you need. It is your responsibility to specifically follow your institution's standard operating procedures (SOPs) and all local, state, and federal guidelines on safe handling and storage of all chemicals and equipment you may use in the labs. This includes determining and using the appropriate personal protective equipment (e.g., goggles, gloves, apron). If you are at any time unsure about an SOP or other regulation, check with your instructor.

Procedure

You will use pieces cut from gloves made of your assigned test material(s). Others in the class will test different chemicals and gloves. After you test your sample(s), you will share the results with others in the class.

1. Put on the appropriate PPE. Your instructor will assign you glove material(s) and chemical(s) to test. Record this information.

2. You will need to do two different samples for each glove material. One sample will be timed at 5-minute exposure and the second at 30-minute exposure. It is important not to confuse the identity of the pieces during this lab. We recommend that you keep track of the glove pieces by placing each into a 250-mL beaker labeled with the type and mass of glove piece it contains, the test chemical, and the duration of the exposure time.

3. Cut out two 1.0- to 1.5-g samples of glove. Mass each sample to ±0.001 g and record.

4. For each sample, follow the test procedure:

    a. Put a sample into a 250-mL beaker. If it does not lie flat, cut it into the fewest pieces required so that all pieces can lie flat in the beaker. Label the beaker.

    b. Barely cover the sample with the test chemical you have been assigned. Start your timing.

    c. Prepare a beaker of rinse water by filling a large beaker about three-quarters full with cold tap water. When the exposure time has elapsed, use forceps to

carefully transfer the sample to the beaker of rinse water. Allow the test sample to stay submerged in the beaker of water for about 1 minute. Pour off the rinse water and replace with fresh tap water. Repeat this rinse step at least twice more.
Use fresh rinse water for each sample.

d. Remove the sample from the rinse water and place it on an appropriate labeled paper towel. Periodically pat the glove sample to help remove excess water and move the sample to a dry place on the towel. Allow to air dry for at least 30 minutes. Mass and record. Allow to air dry for 5 minutes more, mass, and record. Repeat 5-minute intervals of drying and massing until a constant mass at ±0.01 g level is attained. After the air drying is complete (all samples should have dried for about the same amount of time), determine and record the dry mass of the sample.

5. Follow your instructor's directions for cleaning up the test beakers with residual test solution and disposing of the glove samples.

6. Determine the percent change in mass. Rank the glove for durability according to the following rating chart.

$$\% \text{ mass change} = \frac{|\text{final mass} - \text{initial mass}|}{\text{initial mass}} \times 100\%$$

% Change In Mass	Rating
0–10%	(E) Excellent
11–20%	(G) Good
21–30%	(F) Fair
31–49%	(P) Poor
>50%	(NR) Failed

7. Share your results with the rest of the class. How do these results compare with the manufacturers' recommendations for use of the different gloves tested?

Lab: Glove Permeation

In this Lab, you test the permeability of glove material to an acid and base. The procedure you will use is adapted from tests conducted by glove manufacturers to assign ratings to their products.

Safety

In a laboratory setting, you are ultimately responsible for your own safety and for the safety of those around you. In order to more closely resemble an actual workplace, specific safety information is not routinely provided to you for the labs in this book. As discussed throughout the book, many resources are available to assist you in obtaining the information you need. It is your responsibility to specifically follow your institution's standard operating procedures (SOPs) and all local, state, and national guidelines on safe handling and storage of all chemicals and equipment you may use in the labs. This includes determining and using the appropriate personal protective equipment (e.g., goggles, gloves, apron). If you are at any time unsure about an SOP or other regulation, check with your instructor.

Procedure

You will test one glove type but will need three fingers from the glove to complete all tests. Others in the class will test different types of gloves and test chemicals. After you test your sample, you will share and compare your results with others in the class.

1. Put on the appropriate PPE. For each of the three glove fingers, the following setup is required:

    a. Turn the finger inside out and scrub it with soap and tap water and a lab brush. Be careful not to tear the fabric. Rinse thoroughly several times to remove all powder and soap residue.

    b. Slide the finger over the end of a PVC tube. Pull the finger high enough on the tube so it does not slip off, but leave at least 1–2 cm of the tip projecting off the end of the tube. (See Figure 4C-1.) Clamp the finger assembly upright to a ring stand. Label the top of the PVC tube with the test chemical and your initials.

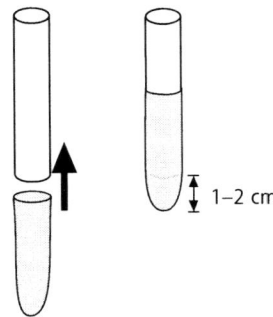

Figure 4C-1: Slide the glove fingers over the PVC tubes.

c. Pour 10 mL distilled water into the finger assembly. Measure and record the pH of the distilled water in each finger.

2. Label two 100-mL beakers with your initials and the name of the appropriate test chemical. Label a third 100-mL beaker with "dist H₂O" and your initials. (This is your control.) Fill each beaker half-full with the appropriate chemical.

3. Set one of the beakers under each finger assembly. Lower the tubes into the beakers so that the finger tip is completely submerged. Ensure that the top of the glove finger is always above the level of the test chemical as shown in Figure 4C-2. Allow the assemblies to stand in the test chemicals for 5 minutes.

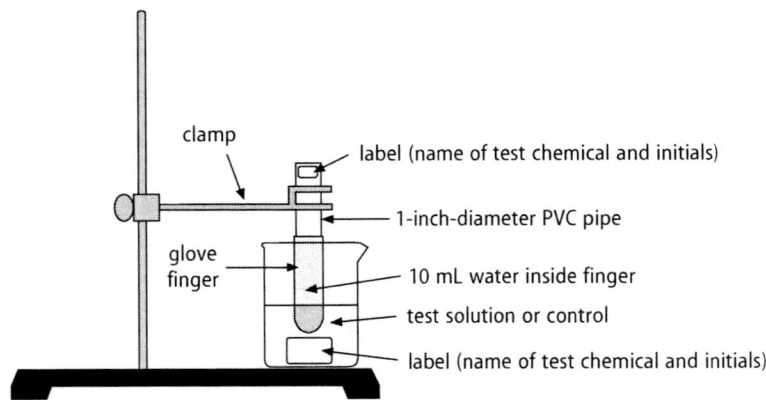

Figure 4C-2: Place the finger assembly in the test solution.

4. Use a clean stirring rod to carefully stir the water inside the tubes before measuring and recording the pH of the water in each tube. You may need to remove the finger from the test chemical beaker to read the pH.

5. Allow the finger assemblies to stand in the test chemical for 25 minutes more for a total of 30 minutes soaking time. Then stir the water inside the tubes, measure the pH, and record.

6. Repeat at 60 minutes of soaking time.

7. To prevent undesired reactions, clean each setup separately. For each, fill a large beaker with tap water. Use the clamp to carefully move the finger assemblies out of the test chemicals into the beaker of water. Still holding onto the clamp, carefully stir the finger assembly around in the water for about 1 minute. Then hold the assembly under running water, rinsing both the outside and inside. Dispose of the glove waste and the beaker of test chemicals as directed by your instructor.

8. Compare your results with others in the class, and discuss class trends and reasons for variations.

Conclusion

What general recommendation can be made about selection of gloves for use with chemicals?

Section 4D: Clothing Selection
Instructor Notes

The objective of this section is for students to become familiar with appropriate clothing selections for laboratory work and determine the chemical resistance of several blends of fabric to various chemicals.

Student Pages

- Section 4D: Clothing Selection—Background
- Section 4D: Clothing Selection—Lab: Fabric Durability

Possible Playout of the Section

Students read the Background and discuss the issue of appropriate laboratory attire. In the Lab "Fabric Durability," students test the outcome of exposing different common clothing fabrics to test chemicals. We recommend that a minimum of four groups be created, each to test a different chemical. Within each group, the work can be subdivided so that all the available fabrics are tested. Because of the extreme hazards associated with handling concentrated acids, we recommend that the instructor conduct the tests using the concentrated acid and a student group be assigned the less concentrated 6 M acid from the pairs as listed in Setup Notes. The effect of concentrated and dilute acids on fabrics can then be safely compared. After collecting data, students should pool class results in order to answer the questions at the end of the Lab.

Safety

The labs in this book do not routinely include special safety precautions for the chemicals or equipment that might be used. As the course instructor, it is expected that you will provide students with access to SOPs, MSDSs, and other resources they need to safely work in the laboratory while meeting all regulatory requirements. Before doing any of the labs from this book or from other sources, we recommend that you regularly review special handling issues with students, allow time for questions, and then assess student understanding of these issues.

Setup Notes

Materials per student
- appropriate personal protective equipment (PPE)
- 2 small beakers

Materials per group
- fabric or clothing made from the following materials:
  - cotton
  - rayon
  - nylon
  - polyester
  - wool
  - leather
- scissors
- dropper (if the assigned chemical is not in a dropper bottle)
- stirring rod
- forceps
- waxed paper or plastic wrap
- masking tape and pen for labels
- stopwatch

Materials per class
- one of the concentrated and less-concentrated acid pairs, such as
  - 18 M H_2SO_4 and 6 M H_2SO_4
  - 12 M HNO_3 and 6 M HNO_3
  - 12 M HCl and 6 M HCl
- moderately concentrated base such as 6 M NaOH
- stain such as 1 M $KMnO_4$ or methylene blue
- common organic solvent such as acetone or toluene

Representative Data

Sample Fabric Testing Results						
	min.	Cotton	Wool	Leather	Polyester	Nylon
18 M H_2SO_4	5	faded, with large hole	slight red stains	slight stain	white spot, shrinking	many holes
	20	fabric eaten away	no further change	no further change	shrunk with holes	shredded
6 M H_2SO_4	5	no change	slight whitening	slightly eaten away	no change	bright white spot, shrunken, holes
	20	no change	bright white	frayed sides	no change	acid melted all
6 M NaOH	5	slightly stained, shrunken, frayed	no change	no change	no change	no change
	20	more shrunken, frayed	slight stain	frayed sides	no change	stain
Methylene Blue	5	dark blue spot	slight blue stain	dark blue throughout	small, blue stain	slight blue stain
	20	whole fabric blue	spot darkened, spread	no further change	no further change	no further change
Acetone	5	no change	no change	slight white discoloration	no change	stained
	20	no change	no change	adhered to glass	no change	no further change

Section 4D: Clothing Selection

Background

This section provides general recommendations for and information about protective equipment. However, you should always follow the protective equipment policy established by your organization's chemical hygiene plan.

If your tastes run to short sleeves, sandals, and long, loose hair, the chemical laboratory is not the place to express yourself through clothing and hairstyles. Choice of hair styles and the appropriate selection of clothing styles and fabrics can minimize hazards due to chemical exposure, machinery, electricity, and fire.

Clothing, Hair, and Jewelry Styles

When working in a laboratory you should always minimize your chemical exposure by wearing personal clothing that is fully covering: long pants, a long-sleeved shirt or a lab jacket, and closed-toe shoes. You will also need to watch out for loose clothing that can fall into chemicals or flames, such as neckties, baggy pants, loose sleeves, and coats (other than lab coats). In selecting clothing styles, consider ease of removal in case of an accident. Selecting appropriate fabric for clothing worn in the laboratory is also important. (See "Protective Clothing" for more details.)

Unrestrained long hair, long beards, and loose jewelry can get caught in equipment, fall into chemicals, or catch fire. Rings, close-fitting bracelets, or absorbent watch bands can trap chemicals next to the skin. Hair spray is extremely flammable and should not be used.

Shoes

Many styles of street shoes do not protect you from chemical and mechanical hazards in the laboratory. You should wear substantial shoes, avoiding clogs, perforated shoes, sandals, and cloth shoes. Steel-toed safety shoes are needed if working with heavy items such as compressed gas cylinders. Shoe covers may be required if you work with especially hazardous materials. In cases where detonation due to static electricity is a hazard, shoes with conductive soles can help prevent buildup of static charge. Conversely, shoes with insulated soles may be needed to protect against electrical shock.

Protective Clothing

Protective clothing such as laboratory aprons and coats may offer you additional protection from injuries or illnesses caused by skin exposure. Other items include jumpsuits, special boots, shoe covers, gauntlets, and splash suits. Such clothing may be required if working with hazardous chemicals, especially if your personal clothing could become contaminated. Occupational Safety and Health Administration regulations (in Title 29 Code of Federal Regulations Subpart I) require each organization to conduct a risk assessment to determine the hazards present and the properties needed for protective clothing.

A wide variety of materials may be used to make chemical protective clothing. As with gloves, each material offers protection against different hazards. Cotton is a good material for laboratory coats, although it reacts rapidly with acids. Plastic or rubber aprons can provide good protection from corrosive liquids, but plastic aprons may accumulate static electricity and therefore pose a danger around flammable solvents. Laboratory clothing made of some types of fabrics can also provide protection against physical or thermal hazards. Generally, chemical protective clothing consists of a barrier material that may or may not be supported on a base fabric. The barrier provides the chemical resistance and the base fabric provides added physical strength. Various combinations of barrier materials and base fabrics result in protective fabrics with properties suitable for different applications.

Some chemical protective clothing is reusable and some is disposed of after a single exposure. The barrier materials used for reusable clothing are often elastomers (such as natural rubber, nitrile rubber, butyl rubber, DuPont Viton®, and neoprene). Some thermoplastics are also used (such as PVC and Teflon®). These materials are often applied to base fabrics such as cotton, polyester, nylon, and blends of these fibers. The barrier materials for single exposure (disposable) clothing are often single or multi-layered thermoplastic films (such as polyethylene and DuPont Saranex®). These barrier materials may be applied to a non-woven fabric base, such as DuPont Tyvek® or other polypropylene materials.

Some of the important properties of chemical protective clothing are listed below:

- Chemical barrier performance: protection from exposure to vapor, liquid, or particulate hazards

- Overall integrity: ability of entire clothing item to prevent inward leakage of gases, liquids, or particles (the clothing design should provide similar performance as the materials from which it is constructed)

- Physical strength and durability: resistance to tears, punctures, and cuts and the ability to maintain barrier performance over the expected period of usage

- Thermal performance: protection from heat or flame (when incident flame or heat contact is expected)

- Static charge: resistance to developing a static charge (when clothing is used in potentially flammable or combustible environments or around sensitive electronic processes)

- Functionality: mobility, dexterity, and comfort while worn

The following table lists some properties of fabrics that affect safety in the laboratory.

Properties of Common Clothing Fibers					
Fiber	Strength	Flammability/ Effect of Heat	Effect of Acids and Alkalis	Effect of Bleaches and Solvents	Static Properties
Cotton	Fairly durable	Flammable; must be treated for flame resistance	Acids cause degradation	Will bleach in most bleaching solutions; generally insoluble in organic solvents	No static problems
Polyester (several types)	Strong, abrasion-resistant	May stick at 440–445°F; may melt above 482°F	Good resistance to organic and inorganic acids at room temperature; moderate resistance at 212°F; good resistance to weak alkalis; disintegrates in concentrated hot alkalis; excellent resistance to acids	Excellent resistance to bleaches and other oxidizing agents	Static problems
Nylon	Exceptionally strong, abrasion-resistant	Melts at 419–430°F; decomposes at 600–730°F; must be treated for flame resistance	Strong oxidizing agents, mineral acids cause degradation; others, loss in tenacity and elongation; soluble in formic and sulfuric acids; hydrolyzed by strong acids at elevated temperatures; substantially inert in alkalis	Will bleach in most bleaching solutions; generally insoluble in organic solvents; soluble in some phenolic compounds.	Static problems
Rayon	Fairly durable	Does not melt; weakens at 300°F	Similar to cotton; hot, dilute, cold concentrated acids disintegrate fiber; strong alkalis cause swelling and reduced strength	Attacked by strong oxidizing agents; not damaged by hypochlorite or peroxide; generally insoluble in organic solvents	No static problems
Acrylic	Good abrasion resistance	Does not melt	Generally good resistance to mineral acids; fair/good resistance to weak alkalis, moderate resistance to strong alkalis at room temperature	Good resistance to bleaches and common solvents	Static problems
Sources: FabricLink Website; *Textile World* Manmade Fiber Chart 1998					

Section 4D: Clothing Selection
Lab: Fabric Durability

In this Lab, you will observe the effects several common laboratory chemicals have on common fabrics used in clothing and shoes.

Safety

In a laboratory setting, you are ultimately responsible for your own safety and for the safety of those around you. In order to more closely resemble an actual workplace, specific safety information is not routinely provided to you for the labs in this book. As discussed throughout the book, many resources are available to assist you in obtaining the information you need. It is your responsibility to specifically follow your institution's standard operating procedures (SOPs) and all local, state, and federal guidelines on safe handling and storage of all chemicals and equipment you may use in the labs. This includes determining and using the appropriate personal protective equipment (e.g., goggles, gloves, apron). If you are at any time unsure about an SOP or other regulation, check with your instructor.

Procedure

1. Put on the appropriate PPE. Divide the available fabric samples as equally as possible among the members of your team. Your team members will pool the results to determine the effects of your assigned test chemical on different fabrics over time. After each group has completed its work, results will be shared as a class.

2. Cut three small fabric swatches (about 3 cm x 3 cm) from the fabric sample you will test. Place one piece of fabric in each of two different beakers. Reserve the third piece as a control. Label one beaker "5 minutes" and the other "30 minutes."

3. Using a dropper, carefully dispense 10 drops of your assigned test chemical onto the fabric square in each beaker.

4. After 5 minutes have elapsed, pour 50 mL tap water into the "5-minute" beaker. Use a stirring rod to gently agitate the fabric in the beaker for about 30 seconds. Decant the liquid, and repeat the water rinse twice more. Use forceps to remove the rinsed fabric, and lay it on a piece of waxed paper or plastic wrap. Label the fabric with the exposure time and test chemical. Record any changes that have occurred.

5. After 20 minutes, rinse, remove, and label the fabric from the "20-minute" beaker as you did in step 4. Compare and record the results of the two different exposure times.

6. Conduct additional fabric testing as needed to complete your group's assignment, and then pool your observations and conclusions with the rest of the class.

7. Work with your team to answer the following questions:

    *   If you knew that dilute acids would be used in the laboratory next week, which fabric would you want to avoid wearing? What if you were going to use concentrated acids?

        The 6 M H_2SO_4 had little effect on anything but nylon, which was severely damaged. Leather and wool showed the least effect with concentrated H_2SO_4. Nylon, polyester, and cotton all were severely damaged.

    *   Review the "Properties of Fabrics" table in the Background. What other tests do you think would need to be done to adequately test fabric for use in the laboratory?

        You would need to consider strength, flammability, and static properties. Also, you would want to test other types of fabrics to provide more options.

Section 4E: Fume Hood Safety
Instructor Notes

The objective of this section is for students to understand the types of fume hoods, when to use them, and how to operate them safely.

Student Pages

- Section 4E: Fume Hood Safety—Background
- Section 4E: Fume Hood Safety—Lab: Evaluating Fume Hood Performance

Possible Playout of the Section

In the Background students learn about the importance of using fume hoods to exhaust potentially harmful vapors and gases and prevent them from being inhaled or introduced into the general air supply of a building. If multiple hoods are being tested at the same time, all hoods must run continuously throughout the entire exercise to minimize airflow disturbances, and groups need to coordinate their tests so all are conducted under a similar set of room conditions. You may want students to compare their results with the manufacturer specifications on the hoods. Typical flow rates are 100 linear ft/min.

In addition to fume hood safety, you may wish to introduce students to the use of respirators by inviting an industrial hygienist to speak on this topic.

Safety

The labs in this book do not routinely include special safety precautions for the chemicals or equipment that might be used. As the course instructor, it is expected that you will provide students with access to SOPs, MSDSs, and other resources they need to safely work in the laboratory while meeting all regulatory requirements. Before doing any of the labs from this book or from other sources, we recommend that you regularly review special handling issues with students, allow time for questions, and then assess student understanding of these issues.

This Lab uses smoke generators to test the efficiency of the fume hoods. The instructor needs to identify people with respiratory problems (such as asthma or allergy sufferers) or others with possible sensitivities to smoke (such as contact-lens wearers) and to ensure that they take proper precautions when a smoke generator is used.

Setup Notes

Materials per group
- laboratory fume hood
- air velocity device (see notes)
- masking tape
- tissue or Kimwipes®
- matches
- smoke generating device (see notes)
- (optional) extra smoke generators
- meterstick or tape measure
- appropriate personal protective equipment (PPE)

Because a large amount of smoke coming out of a fume-hood stack may lead to calls, you may want to inform campus security and/or local fire departments when using smoke generators.

Various instruments are available to measure hood air velocity. One type is a deflecting vane air velocity meter, or velometer. An inexpensive velometer (Vaneometer™) suitable for this Lab is manufactured by Dwyer Instruments, Inc., Michigan City, IN, 219/879-8000, item #480 (under $25 as of this book's printing). Dwyer's device is accurate to ±5% of full scale to 100 linear ft/min and ±10% from 100 linear ft/min to top of scale. Another air velocity device is an anemometer—an electronic device that provides a digital readout. Anemometers are much more expensive than velometers. You may wish to check with your environmental health and safety office and see if they have an air velocity device that you can borrow.

Inexpensive devices for generating smoke include smoke bombs and matches. Smoke bombs of various durations (45 seconds and up) and 10-second smoke matches are available from Regin HBAC Products, Inc., 203/922-0033, *www.regin.com*. Alternatively, your environmental health and safety office may be able to loan you a different type of device such as a chalk gun or a machine that generates smoke.

Section 4E: Fume Hood Safety

Background

This section provides general recommendations for and information about protective equipment. However, you should always follow the protective equipment policy established by your organization's chemical hygiene plan.

Inhalation is the most likely route for chemical exposure. The average person inhales and exhales about 15 times a minute. Each time you inhale, approximately half a liter of air enters your lungs. The air passes through the bronchial system and reaches the alveoli, a network of over 300 million air receptacles that are heavily networked with blood vessels. The red blood cells efficiently transport oxygen from the air into your circulatory system. If the air you breathe is contaminated with hazardous chemicals, this same transport system efficiently carries toxins directly into your bloodstream.

How do you avoid inhaling dangerous quantities of hazardous chemicals? *Prudent Practices* states, "Laboratory fume hoods are the most important components used to protect laboratory workers from exposure to hazardous chemicals and agents used in the laboratory." A laboratory fume hood is a fire- and chemical-resistant enclosure with one opening in the front (face) and a moveable window (sash). (The sashes may open horizontally or vertically.) The laboratory fume hood operates as part of an air supply system for the laboratory as a whole. Large volumes of air are drawn through the face, into exhaust ducts, and out of the building to contain and remove contaminants from the laboratory. Air flow into the hood is generated by exhaust fans. Exhaust fan motors should be installed on the building roof to maintain the exhaust ducts under negative pressure.

The first fume hoods were basically boxes open on one side and hooked up to an exhaust duct. Later, the introduction of baffles improved distribution of exhaust in the hood and airfoils across the face reduced air turbulence. Laboratory fume hoods can be divided into three categories, depending on the pattern of air flow:

- **Non-Bypass**—Air flows into the hood only through the sash opening. As the sash is lowered, the face velocity increases. (See Figure 4E-1.)

- **Bypass**—This design is similar to the non-bypass design as the sash is first lowered (face velocity increases). Once the sash is moved low enough to uncover the bypass opening located on the top front of the hood, air flows through and the face velocity stops increasing and reaches a terminal velocity. (See Figure 4E-2.)

- **Auxiliary Air Fume Hoods**—These hoods were developed in an effort to save energy by minimizing the loss (through the hood) of the building's conditioned air. (Conditioned air is air that has been heated, cooled, filtered, or processed in some other way.) This was done by providing a source of unconditioned (or less conditioned) make-up air directly to the top front of a bypass hood. Ideally, this "auxiliary" air would replace a large portion of the conditioned air previously required for operation. (These hoods have fallen out of favor due to technical problems.)

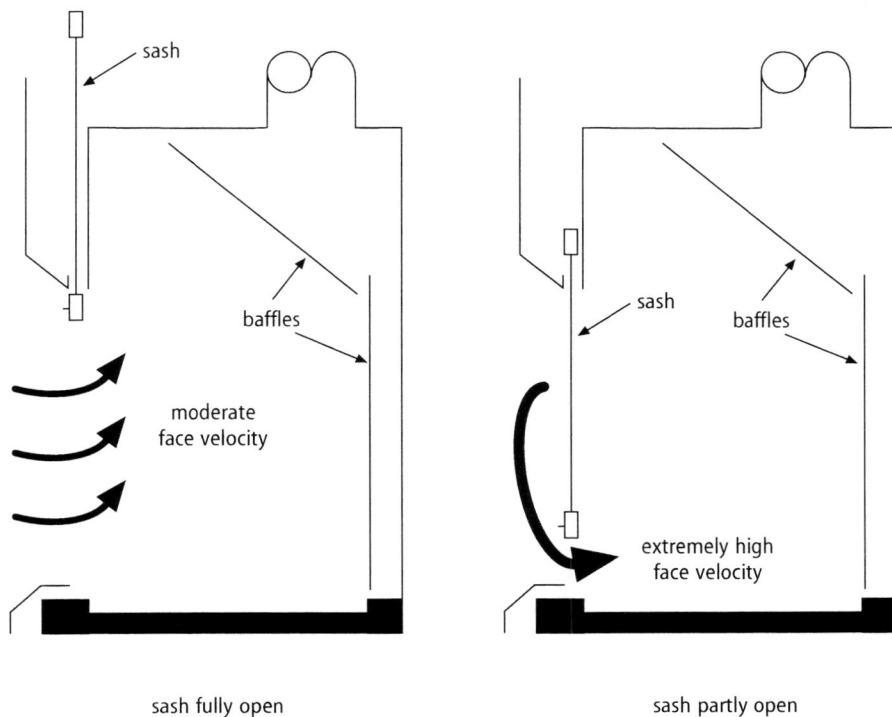

Figure 4E-1: In a non-bypass fume hood, air flows into the hood only through the sash opening.

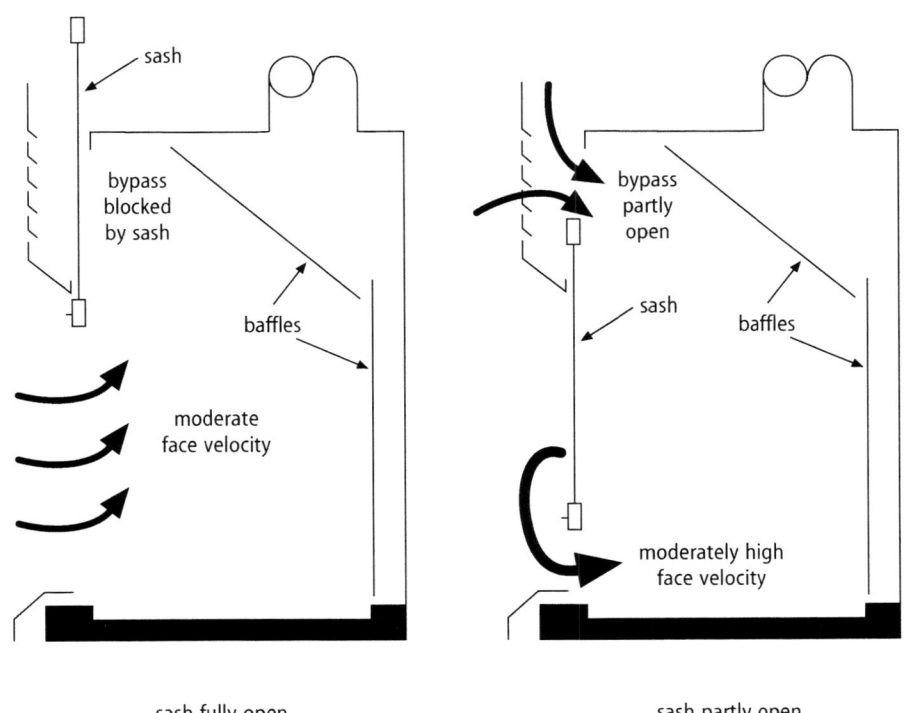

Figure 4E-2: In a bypass fume hood, once the sash is moved low enough, bypass openings on the top front of the hood are uncovered.

Many types of fume hoods are available today, fitting different laboratory configurations and types of work to be done.

- **Benchtop Hoods**—These hoods sit on a laboratory bench and may be bypass or non-bypass. The sash type can be vertical or horizontal or a combination of both. The work surface usually is dished or has a raised lip around the edge to contain spills.

- **Distillation Hoods**—A distillation hood is similar to a benchtop hood, except the work surface is lower to allow room for tall equipment, such as distillation columns, inside the hood.

- **Walk-In Hoods**—This type of hood stands on the floor and is used to contain very tall or large apparatus. It is usually a non-bypass type, and the sash type can be vertical or horizontal or a combination of both. Laboratory personnel should never actually walk into a walk-in hood (despite its name) during an experiment involving hazardous chemicals.

- **Perchloric Acid Hoods**—This type of hood is specially designed for use with perchloric acid and other materials that can form shock-sensitive perchlorates. A water-spray system is used to wash all interior surfaces of the hood system—wash is plumbed to the drain with a special collection system.

When to Use Fume Hoods

The standard operating procedures established by your organization and written in its chemical hygiene plan will state which procedures require fume hoods. Additionally, the MSDS may indicate whether a fume hood is required when handling a chemical.

Evaluating Fume Hood Operation

The efficiency of a fume hood depends on three main factors, as described below. Because different fume hoods have different specifications established for them, you should know these details for the hoods you use.

- **Face velocity** is the average velocity of the air drawn through the face of the hood. The face velocity of the hood greatly affects the hood's ability to contain hazardous air contaminants. Face velocities that are either too low or too high reduce the containment efficiency of the hood. An average face velocity of 100 linear ft/min ($\pm20\%$) is recommended. Hoods with constant air volume (CAV) air supply systems have a specific sash height that will produce the desired face velocity. If you are using a hood with a CAV system, be sure to keep the sash in the proper position. Hoods with a variable air volume (VAV) air supply system can maintain the desired face velocity at various sash heights.

- **Air turbulence** can allow hazardous air contaminants to escape through the face of the fume hood, even where face velocities are appropriate. Smoke tests, often producing colored smoke that is easy to see, are used to visually observe this air movement. A smoke-generating device (available from laboratory supply companies) is placed within the hood and the path of the smoke is observed. Air turbulence can be caused by unnecessary or bulky items obstructing air flow in the hood, rapid arm and body movements inside and outside the hood, and installation of hoods near doors, supply air ducts, or high-traffic areas in the laboratory.

- One of the most meaningful (but time-consuming and expensive) measures of hood performance is to measure **worker exposure** while the hood is being used for its intended purpose. Contact an industrial hygienist for more information regarding quantitative hood performance testing.

Guidelines for Maximizing Fume Hood Efficiency

Many factors can compromise the efficiency of a hood's operation. Most of these are avoidable; thus, it is important to be aware of all behavior that can, in some way, modify the hood and its capabilities. According to *Prudent Practices,* the following precautions should always be observed when using a hood:

- Keep fume hood exhaust fans on at all times.

- If possible, position the fume hood sash so that work is performed by extending the arms under or around the sash, placing the head in front of the sash, and keeping the glass between the worker and the chemical source. The worker views the procedure through the glass, which will act as a primary barrier if a spill, splash, or explosion should occur.

- Avoid opening and closing the fume hood sash rapidly, and avoid swift arm and body movements in front of or inside the hood. These actions may increase turbulence and reduce the effectiveness of fume hood containment.

- Place chemical sources and apparatus at least 6 inches behind the face of the hood. This is necessary because the body acts as an airfoil and can produce vortices that draw the chemical toward the face. In some laboratories, a colored stripe is painted on, or tape applied to, the hood work surface 6 inches back from the face to serve as a reminder. Quantitative fume hood containment tests reveal that the concentration of contaminant can be 300 times higher from a source located at the front of the hood than from a source placed at least 6 inches back. This concentration declines even further as the source is moved farther back toward the back of the hood.

- Place equipment as far to the back of the hood as practical without blocking the bottom baffle slot.

- Separate and elevate each piece of apparatus by using blocks or racks so that air can flow easily and completely around all apparatus.

- Do not use large pieces of equipment in a hood because they tend to cause dead spaces in the airflow and reduce the efficiency of the hood.

- If a large piece of equipment emits fumes or heat outside a fume hood, then have a special-purpose hood designed and installed to ventilate that particular device. This method of ventilation is much more efficient than placing the equipment in the fume hood, and it will consume much less air.

- Do not clutter a hood by using it to store chemicals or unused laboratory equipment.

Section 4E: Fume Hood Safety
Lab: Evaluating Fume Hood Performance

In this Lab, you will become more familiar with the operation of fume hoods in your laboratory and informally evaluate their performance.

Safety

In a laboratory setting, you are ultimately responsible for your own safety and for the safety of those around you. In order to more closely resemble an actual workplace, specific safety information is not routinely provided to you for the labs in this book. As discussed throughout the book, many resources are available to assist you in obtaining the information you need. It is your responsibility to specifically follow your institution's standard operating procedures (SOPs) and all local, state, and national guidelines on safe handling and storage of all chemicals and equipment you may use in the labs. This includes determining and using the appropriate personal protective equipment (e.g., goggles, gloves, apron). If you are at any time unsure about an SOP or other regulation, check with your instructor.

People with respiratory problems (such as asthma or allergy sufferers) or others with possible sensitivities to smoke (such as contact-lens wearers) should alert the instructor and take proper precautions when a smoke generator is used.

Procedure

If multiple hoods are being tested at the same time, all hoods must run continuously throughout the entire exercise to minimize airflow disturbances. Coordinate with the other groups so everyone can test at a given set of room conditions at the same time.

Become familiar with the fume hood.

1. Examine the fume hood at your work area. Does it have airfoils? Baffles? Does it have a vertical or horizontal sash? Is there a sash mark or stop to indicate that someone has determined the best operating position of the sash?

2. Identify the type of fume hood you are working with. Is it a bypass, non-bypass, or auxiliary air fume hood? Does it have a constant air volume or variable air volume air supply system?

3. Note the location of the fume hood with respect to the surroundings. Is it located near doors, windows, air supply vents, cross currents, traffic patterns, or other potential sources of airflow disturbances? Are any chemicals or equipment stored inside the hood?

If there is any clutter inside the hood, ask your instructor if you can move it out. Unnecessary objects can disturb the airflow in the hood.

4. Close the hood door to within 1 inch of the bottom and turn the exhaust blower on. Hold a small piece of tissue or Kimwipe® near the opening. The free end of the tissue should flutter and be drawn into the hood, indicating that air is moving into the hood (the correct direction).

5. Based on your observations so far, do you think the fume hood is well designed? Why or why not? Do you think the location of the fume hood is well planned? Why or why not?

Measure and calculate the average face velocity using an air velocity device.

1. With the sash fully open, measure the dimensions of the open face area of the hood (H and W in Figure 4E-3).

2. Based on the dimensions of the open face area, establish an equally spaced grid representation of the face area, with a maximum section size of about 6 inches x 6 inches. In this example, the grid consists of six squares, as shown in Figure 4E-3.

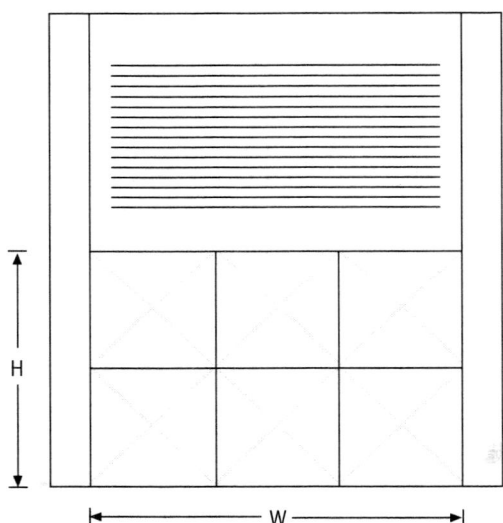

Figure 4E-3: Face of a fume hood in six sections

3. Make sure the fume hood blower is turned on and the sash still fully open. Using a velometer or other air velocity device, measure and record air velocity at the center of each grid square. Do the readings vary widely or are they close in value? Why do you think this is so?

4. Determine the average face velocity in linear feet per minute based on class data. Determine the exhausted air volume in cubic feet per minute using the equation below. Face area is the height (H) times width (W) of the face, as shown in Figure 4E-3.

$$\text{exhausted air volume} \left(\text{ft}^3/\text{min} \right) = \text{average face velocity} \left(\text{ft}/\text{min} \right) \times \text{face area} \left(\text{ft}^2 \right)$$

5. Repeat the procedure to determine the average face velocity and the exhausted air volume for each of the following conditions:

- With the sash moved down (or sideways for horizontal sashes) halfway so the opening size is about half that of the fully opened position.

- With the sash moved down below halfway to decrease the opening size even further.

6. Based on your results, determine the sash position you believe would provide optimal hood performance. Put a piece of tape on the frame to mark this position.

7. Leave the sash in its optimum position. Have someone walk back and forth in front of the hood to disturb airflow while you determine the average face velocity. Does the air disturbance significantly affect the average face velocity?

Test the containment ability of the hood with smoke.

1. Put on the appropriate PPE. Make sure the sash is in its optimum position. Close all doors and windows, and eliminate as much air turbulence outside the hood as possible.

2. Turn the hood exhaust blower on and verify an inward airflow at the face. Place a smoke generating device at the work surface inside the hood, about halfway between the sash and the back of the hood (baffle). Protect the work surface from heat or chemicals.

Clear the fume hood and surrounding area of all flammables before lighting the smoke generator.

3. Light the smoke generator. Observe the movement of the smoke. Is it entirely contained within the fume hood? If not, where is it going? What is the pattern of the smoke's movement?

4. (optional) If extra smoke generators are available, test the containment of the hood under different conditions, such as with windows and doors open, with large objects placed inside the hood, or with deliberately created air turbulence.

Conclusion

Review your results and assess the overall quality and performance of your fume hood. If performance was not satisfactory, how it could be improved? Be sure to consider factors such as the possibility of a malfunctioning blower (belts, fan speed, corroded blades), laboratory air supply, traffic patterns, chemical storage, and clutter.

Chapter 5

Handling Laboratory Equipment Safely

We had a quantity of glassware being cleaned in a solution of sulfuric acid and potassium dichromate. It was left on a counter in the chemistry lab overnight. One of the custodians, apparently curious about what it was, picked up the container. It slipped from his hands and dropped on the lab bench and broke. He didn't bother to clean it up. By the next morning the solution had eaten the side of the cabinet and all the tiles off the floor for a distance of five feet in all directions.

Learning by Accident, page 48

Chapter 5 Objectives

Pertinent Voluntary Industry Standards

- *VIS L2 Standards*—Demonstrate safe handling procedures (e.g., handling cylinders, glassware, moving heavy items).

- *VIS (L2.05):01*—Describe the purpose of, and handle safely, common chemical laboratory equipment.

- *VIS (L2.05):02*—Store, transport, and change compressed-gas cylinders correctly and safely.

- *VIS (L2.05):03*—Choose the proper regulators for gases and other materials under pressure or under vacuum.

- *VIS (L2.05):04*—Manipulate and care for glassware and other apparatus safely, including making connections, cleaning, and storing.

- *VIS (L2.05):06*—Identify common components of electric and electronic circuits that may frequently be maintained by laboratory technicians.

- *VIS (L2.05):07*—Demonstrate a basic awareness of electrical safety and its application to the work environment.

- *VIS L3 Standards*—Use proper techniques for storing and transporting gases and other materials under pressure.

- *VIS (L3.03):08*—Clean laboratory glassware and laboratory equipment made of other materials, using appropriate solvents, detergents, and brushes or cleaning devices.

Objectives of the Chapter

Students will learn

- safe handling procedures for laboratory glassware,

- safe handling procedures for compressed gases, and

- electrical safety in the laboratory environment.

Why Your Students Need to Understand Safe Handling of Laboratory Equipment

Laboratory work involves many types of equipment that can pose chemical and/or physical hazards if not used or maintained correctly. *Prudent Practices in the Laboratory* states, "Many of the accidents that occur in the laboratory can be attributed to improper use or maintenance of laboratory equipment." Through the exercises in this chapter, students gain experience working safely with some of the potentially hazardous equipment they may routinely encounter in the workplace.

Overview of the Chapter

Chapter 5: Safe Handling of Laboratory Equipment contains four sections. General background information and complete instructions for the exercises are provided. The background pages are intended only to provide basic background information on the safe handling of laboratory equipment for your students. We recommend that you extend this background information by having students find and read the pertinent sections of the books in the safety reference library (described in Chapter 1). You may wish to have students briefly report on the information they find during the monthly safety meetings.

Section 5A: Getting to Know Safe Handling of Laboratory Equipment

This section introduces students to the safe handling of laboratory equipment. To initiate interest in the topic, students read and discuss Laboratory Incident Reports concerning accidents that resulted from electrical malfunctions or improper use of laboratory glassware.

Section 5B: Working with Electrical Laboratory Equipment

Students learn about the factors that affect the safety of working with electrically powered laboratory equipment and how to check such equipment for safety. Students conduct an electrical safety inspection of the laboratory.

Section 5C: Working with Compressed Gases

The objective of this section is for students to understand the factors that affect the safety of working with compressed gas equipment and how to check equipment for safety. Students read the Background and gain experience by observing and then trying procedures for working with gas cylinders.

Section 5D: Working with Laboratory Glassware

In this section, students learn about the factors that affect the safety of working with laboratory glassware and how to check such equipment for safety. The demonstrations "Looking at Strains in Glass" and "Prince Rupert Drops" dramatically illustrate the potential hazards of using glassware that has been stressed.

References for Chapter 5

Care and Handling of Laboratory Glassware; Corning: Corning, NY, 1989.

Control of Hazardous Energy (Lockout/Tagout). *Code of Federal Regulations,* 29 CFR 1910.147, 1999. (This is available free online at www.osha.gov.)

CRC Handbook of Laboratory Safety, 2nd ed.; Steere, N.V., Ed.; CRC: Boca Raton, FL, 1988.

Dyer, J. "Demonstration: Prince Rupert Drops, Glass Under Stress." *CHEM 13 News,* November 1988.

Kaufman, J. "Don't Be Shocked." *Chemical Health and Safety.* July/August 1995.

Durst, H.D. et. al. *Experimental Organic Chemistry,* 2nd ed.; McGraw-Hill: New York, 1997.

"Effects of Electrical Energy on Humans," Environment, Safety and Health Manual, Appendix 23B, Lawrence Livermore National Laboratory website. www.llnl.gov/es_and_h/hsm/chapter_23/chap23.html#b (accessed Jan 12, 2000).

"Electric Shock," Casemore Electric website. www.casemore.com/shock.html (accessed Jan 12, 2000).

Ferris, C.D. Electric Shock. In *The Electronics Handbook*, Whitaker, J.C., Ed; CRC: Boca Raton, FL, 1996; p 2197.

General Regulator Instruction Manual; Matheson Gas Products: East Rutherford, NJ, May 1990.

Hofstader, R.; Chapman, K. *Foundations for Excellence in the Chemical Process Industries: Voluntary Industry Standards for Chemical Process Industries Technical Workers;* American Chemical Society: Washington DC, 1997.

"Laboratory Safety," Rice University Department of Electrical and Computer Engineering website. www-ece.rice.edu/~jdw/241/lab0.html (accessed Jan 12, 2000).

Learning by Accident; Mojtabai, F.; Kaufman, J., Eds.; Laboratory Safety Institute: Natick, MA, 1997.

Occupational Exposure to Hazardous Chemicals in Laboratories. *Code of Federal Regulations,* 29 CFR 1910.1450, 1999. (This is available fee online at www.osha.gov.)

"The Physical Effects of Electricity," Bass Associates website. www.bassengineering.com/ (accessed Jan 12, 2000).

"Plug Into Electrical Safety: A Home Electrical Safety Check." National Electrical Safety Foundation: Rosslyn, VA, 1995.

Prudent Practices in the Laboratory: Handling and Disposal of Chemicals; National Academy: Washington, DC, 1995.

Safety in Academic Chemistry Laboratories; American Chemical Society: Washington, DC, 1987.

Safe Handling of Compressed Gases in Containers, 8th ed.; Compressed Gas Association: Arlington, VA, 1991.

Safe Handling of Compressed Gases in the Laboratory and Plant; Matheson Gas Products: East Rutherford, NJ, May 1997.

Section 5A: Getting to Know Safe Handling of Laboratory Equipment
Instructor Notes

This section introduces students to the safe handling of laboratory and electrical equipment. To initiate interest in the topic, students read and discuss Laboratory Incident Reports concerning accidents that resulted from electrical malfunctions or improper use of laboratory glassware.

Student Pages

- Section 5A: Getting to Know Safe Handling of Laboratory Equipment—Laboratory Incident Reports
- Section 5A: Getting to Know Safe Handling of Laboratory Equipment—Background

Possible Playout of the Section

Students read the Laboratory Incident Reports and the Background and discuss the following questions:

1. What are the similarities and differences between the incidents described in the Laboratory Incident Reports?

2. What conclusions can you draw about safety from these incidents?

3. How could each of these accidents have been prevented?

Section 5A: Getting to Know Safe Handling of Laboratory Equipment

Laboratory Incident Reports

Incident Description: Glassware

An advanced chemistry student was attempting to insert a glass tube in a rubber stopper. During the insertion process the glass tube broke and the sharp end of the tube penetrated completely through the end of the thumb. The student had been instructed on the proper procedure and was aware of the procedure, but the student was attempting to save time by taking a short cut.

Learning by Accident, page 63

We had a quantity of glassware being cleaned in a solution of sulfuric acid and potassium dichromate. It was left on a counter in the chemistry lab overnight. One of the custodians, apparently curious about what it was, picked up the container. It slipped from his hands and dropped on the lab bench and broke. He didn't bother to clean it up. By the next morning the solution had eaten the side of the cabinet and all the tiles off the floor for a distance of five feet in all directions.

Learning by Accident, page 48

Incident Description: Electrical

Acetone spilled out of a reaction vessel during the addition of dry ice. It seeped underneath a nearby electronic balance and ignited. The balance was severely damaged, but the fire was extinguished before the reaction vessel broke.

Prudent Practices in the Laboratory, page 112

A friend of mine was cutting grass along the pool side. She was standing on concrete which was wet with water. She had an electric mini grass cutter with a strap attached to her shoulder. As she began to cut the edges of the grass she became practically paralyzed because she was experiencing shock. We removed the grass cutter and then brought her to the emergency room. She experienced numbness and shakes for a while.

Learning by Accident, page 49

Incident Description: Compressed Gas

Several cylinders were placed on the ground in a horizontal position, piled like cordwood. When another cylinder was needed, a worker removed the protective cap from one at the top and rolled the cylinder down the pile toward the ground. It went out of control, and its valve assembly broke off in the process. With gas shooting out of the break, the cylinder, like a rocket, flew about 30 feet through the air. It broke through the rear end of a parked car, shot out the windshield, and continued another 1,500 feet, passing over two houses, and striking the foundation of a third as it landed. Luckily no one was injured.

The Industrial Commission of Ohio, Division of Safety and Hygiene

Section 5A: Getting to Know Safe Handling of Laboratory Equipment

Background

Note: This section provides general recommendations for and information about handling laboratory equipment safely. However, you should always follow the procedures established by your organization's chemical hygiene plan.

"The safety manual? It's under that heap of frayed electrical wires, sharp knives and open containers of who-knows-what."

Laboratory work involves many types of equipment that can pose chemical and/or physical hazards if not used or maintained correctly. *Prudent Practices in the Laboratory* states that "Many of the accidents that occur in the laboratory can be attributed to improper use or maintenance of laboratory equipment."

The most common equipment hazards in the laboratory come from electrically powered devices. Next most common are hazards associated with devices for working with compressed gases and high or low pressures and temperatures. Problems with laboratory equipment may result in exposure to chemical hazards, such as accidental release of ammonia due to a spill or broken container or release of chlorine gas due to leaky valves on compressed gas cylinders. Laboratory equipment poses physical hazards as well, such as slips on water flooding from a cooling system, electrical shock from a stirring motor with a frayed power cord, or cuts delivered by glass shards from an imploding vessel.

Electrically Powered Laboratory Equipment

Electrically powered equipment is used throughout the laboratory for procedures requiring heating, cooling, mixing, and pumping. Electric shock is the major electrical hazard. Also, under certain conditions electrical equipment can serve as an ignition source for flammable or explosive vapors. *You will learn more about electrical safety in Section 5B: Working with Electrical Laboratory Equipment.*

Compressed Gas Equipment

Compressed gas cylinders are often housed in the laboratory to provide sources of oxygen, hydrogen, nitrogen, helium, propane, and other gases. These cylinders pose both chemical and physical hazards. The gases in the cylinders may be flammable, toxic, asphyxiant, or corrosive. Accidental release may occur due to leaky or damaged valves or mix-ups between unlabeled gas lines. The pressure under which the gases are stored and the weight of the cylinders pose physical hazards. *You will learn more about working with compressed gas equipment in Section 5C: Working with Compressed Gases.*

Laboratory Glassware

Glassware requires careful handling and storage procedures to prevent damage. Using chipped or cracked glassware or the wrong type of glassware can lead to breakage, especially at high or low pressure. Vacuum-jacketed glassware and evacuated equipment such as Dewar flasks require extreme care to prevent implosions. *You will learn more about glassware safety in Section 5D: Working with Laboratory Glassware.*

Other Equipment Hazards

Many other types of laboratory equipment can be hazardous, such as water-cooled equipment, vacuum pumps, refrigerators and freezers, stirring and mixing devices, and heating devices. Although these are not discussed in detail in this chapter, a few points about each follow:

- The common use of cooling water in laboratory condensers can create a flooding hazard. Care should be taken to ensure all hoses are properly clamped and maintained.

- Mechanical vacuum pumps, if used in distillation operations, can trap volatile substances and other contaminants. The suction line from the vacuum pump should be fitted with a cold trap, and the output of each pump should be vented to an air exhaust system. The mechanical pump must have proper safeguards, such as belt guards or cages.

- Potential hazards posed by laboratory refrigerators involve accumulation of vapors and spillage of incompatible chemicals. Since it is nearly impossible to adequately vent refrigerators, storage of potentially explosive or toxic substances inside them should be kept to a minimum. All laboratory refrigerators must be special explosion-proof refrigerators. Never store laboratory chemicals in a household-type refrigerator.

- Stirring, mixing, and heating devices pose electrical, burn, and fire hazards. These devices should be used with caution. Never use a hot plate to heat a flammable liquid unless the hot plate is designed for that purpose and has a sealed thermostat. The equipment should be inspected for frayed wires, damaged parts, or other defects.

- Liquefied gases pose the hazard of fire, explosion, cryogenic "burns," and asphyxiation. Personnel handling cryogenic fluids should wear face shields and gloves impervious to the liquid being handled. The work area should be well ventilated.

Section 5B: Working with Electrical Laboratory Equipment
Instructor Notes

The objective of this section is for students to understand the factors that affect the safety of working with electrically powered laboratory equipment and how to check such equipment for safety. Students conduct an electrical safety inspection of the laboratory.

Student Pages

- Section 5B: Working with Electrical Laboratory Equipment—Background
- Section 5B: Working with Electrical Laboratory Equipment—Exercise

Possible Playout of the Section

After reading the Background, students discuss the ramifications of safe use and maintenance of electrical equipment in the laboratory. In preparation for the Exercise, you will probably want to demonstrate the use of the ground monitor and (if used) receptacle tension tester. After students complete the Exercise, you may want to discuss the ramifications of sloppy or incomplete inspections and reports.

Safety

The labs in this book do not routinely include special safety precautions for the chemicals or equipment that might be used. As the course instructor, it is expected that you will provide students with access to SOPs, MSDSs, and other resources they need to safely work in the laboratory while meeting all regulatory requirements. Before doing any of the labs from this book or from other sources, we recommend that you regularly review special handling issues with students, allow time for questions, and then assess student understanding of these issues.

Setup Notes

To complete the electrical safety inspection in the Exercise, students will need to use a ground monitor. Purchasing information is provided below. Optionally, you also may want to provide a receptacle tension tester, but this device is not specifically referred to in the electrical inspection. Groups of students can take turns using the ground monitor. You will also need at least one night-light or lamp.

Ground monitors (also called circuit analyzers) and AC-sensors are available from the Laboratory Safety Institute (508/647-0900, *www.labsafety.org*), and from Flinn Scientific (cat. #SE9095, 800/452-1261, *www.flinnsci.com*). Receptacle tension testers (optional for this exercise) may be available from electrical supply companies. One manufacturer (Daniel Woodhead Company) provides sales information on its website, *www.danielwoodhead.com*.

Section 5B: Working with Electrical Laboratory Equipment

Background

Note: This section provides general recommendations for and information about working with electrical equipment. However, you should always follow the procedures established by your organization's chemical hygiene plan.

Electrically powered equipment is a major part of any chemistry lab. However, electrical energy and the equipment that uses it must be handled very carefully because of the potential dangers. In order to understand the potential hazards posed by electricity and guard against them, you must understand some basic facts about electricity. Electric current always follows the path of least resistance, so an important part of electrical safety involves preventing a person's body from becoming a path of least resistance. Another aspect of electrical safety involves keeping electric current confined to the desired paths, where it cannot come in contact with flammable substances and ignite them.

Electricity, Voltage, and Current

Electric current is the flow of electrons. Charge is measured in coulombs (1 coulomb is equal to the sum of the charges of 6.28×10^{18} electrons), so current is a measure of how many coulombs pass a given point each second. One coulomb per second is defined as an ampere, or amp for short. Direct current (DC) refers to a flow of electrons that moves in one continuous direction, like the current from a battery, which always flows from the negative terminal to the positive terminal. Alternating current (AC) is current that moves in one direction for a certain amount of time, then reverses itself in an alternating pattern. The number of times the current reverses direction in 1 second is known as frequency and is measured in hertz (Hz). In the US, household current has a frequency of 60 Hz.

A conductor is a material that contains mobile charged particles. Metals are good conductors, because the valence electrons of metal atoms are loosely attached and can move easily. Batteries, salt water, the earth, and the human body are also good conductors, as they contain electrolytes. An insulator, on the other hand, is a material whose valence electrons are tightly bound and do not move easily. Rubber, wood, glass, and many plastics are good insulators. Resistance is a measure of a material's ability to oppose the flow of electric current; a good conductor has a low resistance, and an insulator has a nearly infinite resistance. Resistance is measured in units called ohms (Ω).

Voltage is a measure of the electric potential at a particular reference point. It can help to think of voltage as an electric "pressure." If we compare voltage to the water pressure at a water faucet, electric current corresponds to the flow of water coming out of the faucet, and resistance is inversely proportional to the size of the opening allowed by adjusting the faucet. If the water pressure remains constant and the faucet handle is turned to decrease the opening size (increase the resistance to flow), then the flow of water decreases. Just as the water in this example always flows out of, or away from, the faucet (from a higher to a lower pressure), the direction of current is always from a higher voltage to a lower one. The simple relationship between current, voltage, and resistance is summarized in Ohm's Law, which states that the voltage difference between two points on a conductor equals the current multiplied by the resistance of the conductor: Voltage = Current x Resistance. Voltage is measured in units called volts (V). Typical home and academic electrical outlets are 120 V. Large appliances such as clothes dryers and stoves require 240-V outlets. Some industrial outlets will have voltages as high as 480 V. Each voltage outlet has a unique design to prevent mistakes. In other words, you can't plug a 120-V plug into a 240-V outlet.

Electric Shock

Electric shock is always a potential hazard when working with electrically powered equipment, machinery, and devices. Electric shock occurs when electric current flows through a person's body. The severity of an electric shock depends on a number of variables including the amount of current (determined by the voltage and the body's resistance), the path the current takes through the body, the type of current (DC or AC), and duration of contact. These variables interact in complex ways to produce a variety of potential injuries.

Because the human nervous system is partly electric in nature, people are sensitive to very small currents. The median threshold of perception (the smallest current that can be felt) on the hand is about 5 milliamperes (mA) DC and 1 mA at 60Hz AC. (A milliampere is 1/1,000 ampere.) As the amount of current increases, voluntary control of muscles in the path of the current becomes increasingly difficult until finally the victim "freezes" to the circuit. The level of current at which a person can just release an electrode is known as "let-go" current. The average let-go current for healthy adults is about 16 mA; however, let-go currents as low as 5 mA have been measured.

For a given voltage, the amount of current that passes through a body depends on the body's resistance. The average person's body resistance is between 500,000 and 1,000,000 ohms, but resistances can fall far below this range depending on a number of physical factors. Moisture and electrolytes in and on the body can greatly affect how good an electric path a person is. Very dry skin has a higher resistance, while skin soaked with salt water has a lower resistance. Since current is inversely proportional to resistance, the higher the resistance offered by a human body, the lower the amount of current exposure. Thus, the best strategy for reducing the severity of an accidental shock is to increase the resistance of the body—make sure your hands, the equipment's power switch, and the floor are dry before touching any electrical appliance or wire. Stop using a piece of equipment and immediately inform your instructor or supervisor if you ever feel even a small tingle of electric current when touching the equipment.

Voltage becomes a more significant factor at lower body resistances. Fatal shock (electrocution) can result from currents associated with voltages as low as 60 V AC. (Normal household voltages are 120 V and 240 V AC.) Low voltages are not normally fatal, but severe burns are possible.

The path that an electric current takes through the body is probably the most important determinant of whether a shock will be lethal. The most severe type of electric shock occurs when the current passes through either the chest cavity or the trunk of the body—these are considered critical paths. Current flowing through the heart causes it to contract and stay contracted until the current stops flowing. When current flows through the diaphragm and other muscles that enable the lungs to expand and contract, the lungs cannot function. When a current stops flowing through the heart, the heart may not resume a normal beat pattern—ventricular fibrillation (repeated, uncoordinated contractions of the heart ventricles) may occur. In each of these cases, oxygenated blood cannot reach the brain and other parts of the body. Brain damage due to lack of oxygen occurs within 5 minutes.

The following table describes the effects of varying levels of critical-path current on the human body, taking into account some of the variables discussed above. The values should be considered only approximations because of the complexities involved in determining the physiological effects of electric shock. The table below is based on information in the *CRC Handbook of Laboratory Safety*. You may see different values in other sources.

Effects of Electric Current in a Human				
Effect	Direct Current (milliamperes)		Alternating Current (milliamperes)	
	Men	Women	Men	Women
Slight sensation on hand	1	0.6	0.4	0.3
Perception threshold, median	5.2	3.5	1.1	0.7
Shock (not painful, muscular control not lost)	9	6	1.8	1.2
Painful shock (muscular control lost by 0.5%)	62	41	9	6
Painful and severe shock (breathing difficult, muscular control lost by 99.5%)	90	60	23	15
Possible ventricular fibrillation:				
3-second shocks	500	500	675	675
Short shocks (T in seconds)			$116/\sqrt{T}$	$116/\sqrt{T}$
Capacitor discharges (in watt-seconds)	50	50		
Adapted from *CRC Handbook of Laboratory Safety*, p. 525				

two-pole
non-polarized

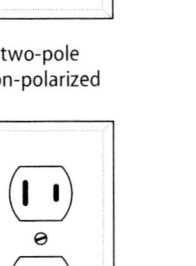

two-pole
polarized

two-pole
three-wire
grounding

Figure 5B-1: Three kinds of receptacles

All electrical equipment in the laboratory must be installed and maintained in accordance with the provisions of the National Electric Code of the National Fire Protection Association (NFPA). The state and local codes that apply to your workplace may be more stringent and/or contain special provisions. Before participating in an experiment involving electrical equipment, make sure that you are aware of all applicable safety issues and have been briefed on potential problems. If you find any electrical hazards in the workplace, be certain that the appropriate person is notified.

Electric Outlets (Receptacles, 120 V only)

Electric outlets (receptacles) can be in one of three configurations: two-pole non-polarized, two-pole polarized, and two-pole three-wire grounding. (See Figure 5B-1.) The purpose of polarized appliance plugs, in which the "neutral" prong is larger than the "hot" (energized) prong, is to force the user to insert the plug into a polarized receptacle so that the hot and neutral prongs of the plug line up correctly with the hot and neutral prongs inside the receptacle. The neutral prong is electrically grounded, meaning that it makes an electric connection with the earth. Be cautious with non-polarized two-pronged plugs, because both prongs are the same size. Inserting a non-polarized plug with the prongs reversed may create an electrical hazard. This hazard is increased if the equipment has a metal housing.

Besides polarized plugs, another design intended to prevent electric shock is the three-wire grounding outlet. A three-wire grounding outlet contains slots for two polarized prongs and a round metal grounding pin, which connects to a grounding wire inside the receptacle. The grounding pin is the longest prong on a three-pronged plug, which means that it is the first thing to make contact inside the receptacle and the last thing to release contact when the plug is removed.

Prudent Practices recommends that all 110- to 120-V electric outlets in laboratories should be of the three-wire grounding type to accommodate three-pronged plugs. The three-wire grounding plug is a valuable safety device on equipment with metal cases because it provides a path to ground that has lower resistance than a path through a human body. Thus, if the hot wire should accidentally work loose and come into contact with the metal casing, the current would preferentially travel to ground through the ground wire, not through a person touching the casing. Three-pronged plugs are not used for all equipment, however. Most new electrically powered tools do not have metal cases, and they are double-insulated. Thus, they don't need a ground wire because even if the hot wire should accidentally work loose and come into contact with the casing, the plastic would not conduct electricity.

Even when equipment users are careful to plug in the equipment correctly, electrical hazards can still exist. For example, sometimes receptacles themselves are wired incorrectly. Hot and neutral wires may be accidentally reversed or the neutral wire left open, or ground wires may not actually be grounded. The outlet receptacles may appear to be wired correctly, but the only way to be certain is to test with a ground monitor or circuit analyzer. These simple devices plug directly into the outlet. A pattern of lights indicates whether the outlet is correctly grounded or not.

In addition, over time, the prongs inside receptacles (the parts that grip the plug prongs) can become loose. Receptacles with loose prongs should be replaced.

Lockout/Tagout

From time to time, all laboratory equipment must be serviced and maintained. OSHA regulations (29 CFR 1910.147) require that lockout/tagout procedures be used to prevent the accidental energizing of machines or equipment being serviced, maintained, or adjusted. Lockout/tagout may involve various procedures, depending on the organization. Generally, a locking device is placed over the power cord to prevent it from being inserted into an outlet and a warning tag is attached to the power cord. These measures are intended to indicate that the equipment is safe to be worked on and prevent all but properly trained and qualified personnel from energizing the equipment. Unfortunately, even the best lockout/tagout procedures can be intentionally defeated, posing a serious hazard. In the workplace, you will be responsible for understanding and following the lockout/tagout procedures at your institution.

At times you may be expected to make adjustments or minor repairs to electrically powered laboratory equipment yourself. Always unplug equipment (unless it must be on in order to be adjusted) and follow any required lockout/tagout procedures. If you must handle equipment that is energized, be certain that your hands are dry, and wear nonconductive gloves and shoes.

Shock Protectors

Electric shocks can be more serious in certain locations, such as where people can contact heating radiators, water pipes, electric heaters, and water in sinks. Touching one of these objects and a faulty electric appliance at the same time can result in a shock and possibly electrocution. Many of the electrocutions that occur every year could be prevented with the installation of a device called a Ground Fault Circuit Interrupter (GFCI). A ground fault is a situation in which current is flowing through something other than normal electric wiring paths—for example, your body.

GFCIs continually monitor electric current flowing within the circuit for any differences in current flowing through the hot lead and the neutral lead. If the magnitude of current flowing into the circuit differs by even a slight amount from the magnitude of current returning, the GFCI quickly shuts off the current flowing through that circuit, preventing serious injury from electric shock. (However, a person may receive a very slight shock during the time that it takes for the GFCI to cut off the electricity.) A GFCI should be used in any area where water may come in contact with electric devices.

Figure 5B-2: GFCI wall outlets can be installed in place of standard outlets.

Three types of GFCIs are available: circuit breaker, wall receptacle, and portable plug-in. A GFCI circuit breaker can be installed on circuit breaker electric panels to protect against electrocution, excessive current leakage, and overcurrent for an entire branch circuit. Circuit breaker-type GFCIs should be installed only by qualified electricians. GFCI wall outlets (see Figure 5B-2) can be installed in place of standard outlets to protect against electrocution for just that outlet or for a series of outlets in the same branch. This type of GFCI can be installed by persons who are familiar with electric wiring practices and follow the instructions for that device. Plug-in GFCIs can be plugged into wall outlets where appliances will be used. These GFCIs do not require any special knowledge or equipment to install. Both two-conductor and three-conductor receptacle outlets can be protected with plug-in units. The National Electric Code requires these devices in new construction for receptacles near sinks, near water sources, outdoors, and in other hazardous locations.

Fuses and Circuit Breakers

Overloaded circuits produce excessive heat and may damage insulation, possibly leading to fires. Overloading means that the appliances and lighting in the building regularly demand more electric current than the electric system can safely deliver. Fuses and circuit breakers are safety devices located on electric panels to prevent overloading of electric circuits. They interrupt the electric current if it exceeds the safe level for some portion of the electric system. If the demand for electric current exceeds the safety level, a fuse opens ("blows"), interrupting the circuit, and must be replaced in order for the circuit to be reconnected. A circuit breaker works a bit differently. It "trips" its switch to break the circuit, and the circuit is reconnected by closing the switch manually. There are two main types of circuit breakers. One has a control handle that swings all the way to "OFF" when it is tripped. The other has an intermediate position midway between "ON" and "OFF." Both types of circuit breakers must be reset with the hand control after the problem has been eliminated. The first type should simply be moved back to "ON," while the second should be moved first to "OFF" and then to "ON."

A tripped fuse or circuit breaker may be an indication that an unsafe condition exists within the circuit. Therefore, before simply replacing the fuse or resetting the breaker, find out which piece of equipment caused the problem and remove it from the circuit. If an overload condition exists, relocate some equipment items to a different circuit.

Sparking from Electric Motors

Electric shock is only one of the potential hazards associated with electric equipment in the laboratory. A series-wound or Universal AC type of electric motor can pose a sparking hazard around volatile flammable materials. Many pieces of laboratory equipment use this type of motor, including vacuum pumps, mechanical shakers, stirring motors, and magnetic stirrers. Laboratory motors, according to Underwriters Laboratories, Inc. (UL) and OSHA, should be rated for laboratory environments. Series-wound motors, unless shielded, are not approved. For work in the presence of flammable vapors, nonsparking induction motors or air motors must be used. Kitchen appliances such as mixers and blenders often have series-wound motors. Therefore, do not use these appliances where flammable vapors may be present. If household equipment with a series-wound motor, such as a vacuum cleaner or electric drill, must be used in the laboratory, make sure that no flammable vapors are present. In general, good safety practice requires that you not use any motorized equipment around flammable vapors unless you are certain the motor is rated for such an environment. Do not use a commercial refrigerator to store organic solvents or other flammable materials; use only refrigerators that are specially rated for these materials. Flammable materials storage and explosion-proof refrigerators feature spark-proof interiors. The motors, thermometers, and wiring are mounted on the outside of the refrigerator when possible, and if mounted on the inside, they are shielded with conduit or sealed to prevent sparking.

Vapor-proof and explosion-proof electric outlets and switches are also available to reduce hazards in environments where flammable materials are present. The National Fire Protection Association (NFPA 70) provides guidelines for determining what environments require the use of these devices.

Potential Electrical Hazards and Their Symptoms

The following are common electrical problems and their symptoms:

- Power outages—fuses need replacement or circuit breakers need resetting frequently.

- Overrated panel—electric panel contains fuses or circuit breakers rated at higher currents than the current capacity (ampacity) of their branch circuits, sometimes called "overfused."

- Dim/flickering lights—lights dim when devices start.

- Arcs/sparks—bright flashes or showers of sparks occur anywhere in your electric system.

- Sizzles/buzzes—unusual sounds come from the electric system.

- Overheating—parts of your electric system, such as switch plates, wall outlet covers, cords and plugs, are warm. These should never be hot, painful to touch, or discolored from heat.

- Permanently installed extension cords—cords are incorrectly used to extend the wiring system permanently, instead of being used temporarily to connect an appliance to an outlet.

- Loose plugs—outlet attachment plugs wobble or easily pull out of a wall outlet.

- Damaged insulation—insulation is cut, broken, or cracked.

Section 5B: Working with Electrical Laboratory Equipment

Exercise

Conduct an electrical safety inspection of the laboratory using the following checklist and test equipment provided by your instructor. Once completed, discuss findings as a class. Prepare a class report with recommendations.

Cords

Q. Are telephone, appliance, and other cords placed out of the flow of traffic?

Cords stretched across walkways may cause someone to trip.

- Arrange laboratory furnishings and equipment so that outlets are available for appliances, computers, and instruments without their cords hanging where people might trip or become entangled.
- If you must use an extension cord, place it on the floor against a wall where people cannot trip over it. Tape it in place.

Q. Are cords placed so they are not beneath laboratory furnishings, floor coverings, and equipment?

Objects resting on cords can crush them. Electric cords that run under floor coverings can become damaged, overheat, and cause a fire.

- Remove cords from underneath laboratory furnishings, floor coverings, and equipment.
- Replace damaged or frayed cords.

Q. Are cords attached to the walls, baseboards, or any other surface with nails or staples?

Nails or uninsulated staples can damage cords, posing fire and shock hazards.

- Disconnect power from the device(s), and then remove nails and staples from the cords.
- Check wiring for visible signs of damage, and use an AC sensor to make sure the wires inside the cord are unbroken. To use the AC sensor, plug in the appliance or instrument (it is not necessary to turn it on), and move the tip of the sensor along the entire length of the cord, from the plug to the device. The light in the tip of the sensor should glow continuously (or intermittently in the case of round cords containing twisted wires). If the light goes out permanently at any point along the cord, the circuit is broken at that point, and the cord should be replaced.
- Use tape if necessary to attach cords to walls or floors.

Q. Are electric cords in good condition (not frayed or cracked)?

Damaged cords may cause a shock or fire. Keep in mind that rubber-covered cords can be eroded by organic solvents and ozone.

- Check for broken or dried-out wiring jackets, missing fiber caps (on older plugs), and frayed wire on equipment. Replace frayed or cracked cords.
- Do not use frayed electric cords.

Q. Are extension cords used on a temporary basis only, not for permanent installations?

Extension cords are not intended for permanent use. OSHA considers extension cords to be temporary wiring and limits their use as described in 29 CFR 1910.305. They can pose a fire hazard.

- Move equipment closer to outlets so extension cords are not necessary on a permanent basis.
- Add necessary permanent electrical service.

Q. Do extension cords carry no more than their proper load, as indicated by the ratings labeled on the cord and the equipment?

Overloaded extension cords may cause fires. If the cord feels hot, it is carrying too high a load.

- Replace No. 18 gauge cords with No. 16 gauge cords. (Smaller gauge numbers indicate larger size and higher capacity.) Older extension cords using small (No. 18 gauge) wires are only rated for about 7 amps, while No. 16 is rated at about 10 amps.

- Change the cord to a higher-rated one, or unplug some equipment if the rating on the cord is exceeded because of the power requirements of one or more pieces of equipment being used on the cord.

- Use an extension cord having a sufficient current rating, if an extension cord must be used.

Light Bulbs

Q. Are light bulbs the appropriate size and type for the lamps or fixtures?

A bulb of too high wattage or the wrong type may lead to fire through overheating. Ceiling fixtures, recessed lights, and "hooded" lamps will trap heat.

- Replace with a bulb of the correct type and wattage. (If you do not know the correct wattage, contact the manufacturer of the fixture.)

- Place halogen lamps away from paper or fabric. These lamps become very hot and can cause a fire.

Shock Protectors

Q. Have you tested your GFCIs to be sure they still offer protection from fatal electric shock?

Sometimes GFCIs can malfunction in such a way that even though they still provide power, they no longer provide shock protection. Be sure your GFCI still provides protection from fatal electric shock.

- Test monthly. First plug a night light or lamp into the GFCI-protected wall outlet (the light should be turned on), then depress the "TEST" button on the GFCI. If the GFCI is working properly, the light should go out. If not, have the GFCI replaced. Reset the GFCI to restore power.

- If the "RESET" button pops out but the light does not go out, the GFCI has been improperly wired and does not offer shock protection at that wall outlet. Contact a qualified electrician to correct any wiring errors.

Fuses/Circuit Breakers

Q. Is the circuit breaker panel free from obstruction?

A circuit breaker even partially obstructed is difficult to reach quickly in an emergency.

- Move laboratory furnishings or equipment that limit access to the panel or prevent the panel door from opening completely. OSHA requires a minimum free space of 30 inches in front of circuit breaker panels (29 CFR 1910.303). If the panel is locked, keys to unlock it must be readily available.

Q. Are all of the circuit breakers in the panel(s) clearly labeled?

When a circuit breaker trips, it is important to know which circuit caused the problem so that the problem can be corrected. Also, labeling is helpful in cases when only one particular circuit needs to be shut off.

- Label circuits legibly and descriptively, using a pen, not a pencil.

Receptacle Outlets and Switches

Q. Are any outlets or switches unusually warm or hot to the touch?

Unusually warm or hot outlets or switches may indicate that an unsafe wiring condition exists.

- Unplug cords from these outlets and do not use switches.

- Have a qualified electrician check the wiring as soon as possible.

Q. Do all outlets and switches have cover plates so that no wiring is exposed?

Exposed wiring presents a shock hazard.

- Add a cover plate.

Q. Do the receptacles grip plugs firmly?

If the prongs inside a receptacle do not grip the prongs of the plug firmly, electric appliances may become unplugged suddenly and the plug may become quite hot due to vibrations.

- Use a receptacle tension tester (if you have one) to determine whether the prongs in the receptacles grip the prongs of plugs strongly enough. While the design of tension testers varies, typically you insert the device in a receptacle, and an indicator on the device tells you whether the prongs in each contact hold the prongs of plugs firmly. The firmness of the grip is rated on a scale (such as 1–5), with 1 being the loosest grip and 5 (or the highest number) being the firmest grip. For readings of 1–3, the receptacle should be replaced. Readings of 4–5 are satisfactory.

Q. Is the receptacle wired correctly?

- Use a ground monitor (also called a circuit analyzer) to determine whether the receptacle is wired correctly. Ground monitors differ, but in general, if you plug one into a receptacle, a pattern of lights on the device will indicate whether the wiring is correct or whether any contacts are reversed or open. The device should have a key permanently attached or printed on it that tells what each pattern of lights means.

Electric Equipment

Q. Are small electric appliances, such as hot plates, unplugged when not in use?

Even an appliance that is not turned on, such as a hot plate, is potentially dangerous if left plugged in. For example, if it falls into water in a sink while plugged in, it could electrocute someone.

- Request that GFCIs be installed near sinks to protect against electric shock. For more information, see the section on GFCIs.
- Unplug all small appliances when they are not in use.
- Never reach into water to pull out an appliance that has fallen in without being sure the appliance is unplugged or the appropriate circuit breaker is cut off.

Q. Are equipment and power tools equipped with a three-pronged plug or marked to show they are double-insulated?

These safety features reduce the risk of an electric shock.

- If you must use a three-pronged adapter to connect a three-pronged plug to a two-pronged receptacle, make sure it is properly connected and grounded.
- Consider replacing old equipment that neither has a three-pronged plug nor is double-insulated.

Q. Are power tool guards in place?

Power tools used with guards removed pose a serious risk of injury from sharp edges or moving parts.

- Replace guards that have been removed from power tools.

Q. Are the grounding features of any three-pronged plugs being properly used—that is, are the grounding pins still present?

The third prong is there because the appliance must be grounded to prevent electric shock. The grounding pin is the first thing that touches the receptacle and the last thing that leaves it.

- Replace appliances or cords whose grounding pins have been removed.
- Don't cut off or trim prongs on polarized plugs.

Q. Are two-pronged non-polarized plugs inserted so that the ribbed side of the cord is connected to the wider (neutral) side of the receptacle?

Inserting two-pronged unpolarized plugs backward can create a shock hazard.

- Insert all plugs correctly.

Section 5C: Working with Compressed Gases
Instructor Notes

The objective of this section is for students to understand the factors that affect the safety of working with compressed gas equipment and how to check equipment for safety. Students read the Background and gain experience by observing and then trying procedures for working with gas cylinders.

Student Pages

- Section 5C: Working with Compressed Gases—Background

Possible Playout of the Section

Students read the Background and discuss the hazards associated with high-pressure compressed-gas cylinders, the properties of compressed gases that represent hazards and the importance of knowing these properties before using a given gas.

Use an empty compressed gas cylinder to demonstrate the correct procedure for moving a cylinder, selecting and attaching a regulator, and connecting the cylinder to a laboratory apparatus. Oversee students practicing these tasks. You may want them to use the "Sample Procedure for Working with Gas Cylinders" provided in the Background. While students are practicing, classmates should observe and then provide constructive feedback.

We suggest using a variety of sizes and types of empty compressed gas cylinders to provide students with practice using different types of cylinders they may encounter in the workplace. After students practice with the empty cylinders, we suggest that you use a full cylinder to demonstrate checking the cylinder and connections for leaks using soapy water, dispensing gas, and shutting off the gas. As you do so, have students observe the cylinder-pressure and delivery-pressure gauges on the regulator. If opportunity allows, students can dispense a given quantity of gas under your supervision. You may also wish to discuss that several types of gas regulators exist and that the type used depends on the gas being regulated.

Safety

The labs in this book do not routinely include special safety precautions for the chemicals or equipment that might be used. As the course instructor, it is expected that you will provide students with access to SOPs, MSDSs, and other resources they need to safely work in the laboratory while meeting all regulatory requirements. Before doing any of the labs from this book or from other sources, we recommend that you regularly review special handling issues with students, allow time for questions, and then assess student understanding of these issues.

Setup Notes

Materials
- empty high-pressure compressed-gas cylinder(s), originally containing compressed air, with appropriately fitting regulator(s)
- hand truck(s) for cylinders
- flat-faced or adjustable wrench
- restraining strap(s) for securing cylinder(s) to a laboratory bench

- personal protective equipment (PPE): goggles, gloves, and protective shoe coverings
  The gloves should be clean and free of dirt, oil, and grease. They should also be made of an impermeable material that protects against cold and be loose-fitting so they can easily be removed in case something spills into the gloves.

- full cylinder, already connected to regulator and tubing
- soap-water solution (1:10 solution of dish soap and water)

For safety reasons, we recommend using empty compressed-air cylinders for the labs. This prevents the potential hazard of residual gas escaping from an empty cylinder that originally held flammable, toxic, or asphyxiant gases. Take all necessary precautions to ensure that the empty cylinder contains nothing but residual air. You may be able to obtain empty compressed-air cylinders from a local specialty gas supplier, such as Air Products and Chemicals at *www.airproducts.com/* or the supplier of specialty gases for your institution. These cylinders will have CGA valves 346 or 590.

Section 5C: Working with Compressed Gases

Background

Note: This section provides general recommendations for and information about working with compressed-gas cylinders. However, you should always follow the procedures established by your organization's chemical hygiene plan.

Compressed-gas cylinders and gas lines provide sources of oxygen, hydrogen, nitrogen, helium, propane, and other gases. Compressed-gas cylinders pose both chemical and physical hazards. The gases may be flammable, toxic, chemically active, asphyxiant, or corrosive. Accidental release may occur due to leaky or damaged valves or mix-ups between unlabeled gas lines. Because of the pressure under which compressed gases are stored, gas tanks must be considered potential torpedoes. (Compressed-gas cylinders have 2½ times the propulsion ratio of the space shuttle!) Also, the weight of the cylinders can result in injury if they fall on someone or are lifted improperly.

According to *Prudent Practices in the Laboratory,* "Precautions are necessary for handling the various types of compressed gases, the cylinders that contain them, the regulators used to control their flow, the piping used to confine them during their flow, and the vessels in which they are ultimately used."

Identification of Contents

Before using any compressed-gas cylinder, you must be certain of the contents. All cylinders should be identified clearly with permanent markings stenciled or stamped on the cylinder itself or with a durable label that cannot be removed from the cylinder. Do not depend on color coding, as cylinder colors vary with suppliers. Labels on caps are also not dependable because caps can be interchanged.

All gas lines leading from a compressed-gas supply should be clearly labeled with the identity of the gas, the laboratory served, and emergency telephone numbers. The labels should also be dated and color-coded as follows: a yellow background with black letters is used for flammable, toxic, or corrosive substances such as acetylene, methane, and ammonia; a green background with black letters is used for inert gases such as helium and nitrogen.

Handling and Use

Compressed-gas cylinders are equipped with valves at the top where the gas is released. Standard markings are used on the sides of each cylinder to indicate important information about the cylinder.

The functions of the valve parts and cylinder markings (shown in Figure 5C-1) are as follows:

- cylinder cap: protects the valve

- valve: where the gas is released

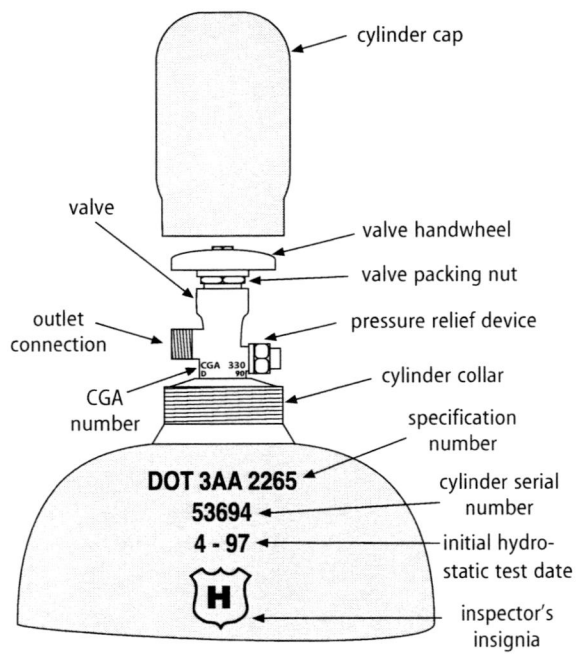

Figure 5C-1: The parts of a compressed-gas cylinder

- valve handwheel: used to open and close the valve

- valve packing nut: contains a packing gland and packing around the stem

- pressure relief device: permits gas to escape if pressure builds to an unsafe level

- outlet connection: connects to equipment that regulates pressure and/or flow

- CGA number: type of standard cylinder-valve outlet connection used

- cylinder collar: holds the cylinder cap

- outlet cap: protects the valve threads from damage and keeps the outlet clean (not shown; not used universally)

- specification number: signifies that the cylinder conforms to Department of Transportation specifications

- cylinder serial number: identifies the individual cylinder

- date of initial hydrostatic testing: indicates the month and year of the cylinder's initial hydrostatic testing

- original inspector's insignia: for conducting hydrostatic and other required tests to approve the cylinder under DOT specifications

The DOT requires periodic testing to confirm that a cylinder can operate safely at the appropriate service pressure. Either hydrostatic or ultrasonic testing may be used. In hydrostatic testing, the cylinder is filled with water, placed in a vat of water, and pressurized to 5/3 of its service pressure. As the cylinder is pressurized, it expands, displacing some water from the vat. When the pressure is released, the cylinder returns to its original size (or very close to it). If the expansion under pressure and the subsequent return to normal size are within required limits, the cylinder can be used and is stamped with the test date. For most types of cylinders, if 5 years have passed since the stamped test date, the cylinder cannot be refilled until it is re-tested. For some cylinders, 10 years may pass before re-testing is required. In this case, a star appears after the test date. The ultrasonic test is relatively new. When it is used, the letters UT appear after the test date.

The Compressed Gas Association (CGA) has devised standard cylinder-valve outlet connections to ensure a gas-tight seal and to minimize the possibility of connecting a gas to a system not designed for that gas. Gas suppliers have detailed information on the equipment assemblies to be used with different gases. To minimize hazards, use only CGA standard combinations of valves and fittings; avoid assemblies of miscellaneous parts. The following table shows five categories of gases, the CGA number for each category, and some examples of gases in those categories.

Categories of Gases, Cylinder Valve Outlets and Connections, and Gases				
Gas Category	CGA Valve Outlet and Connection	Examples of Gases		
Corrosive	330	Boron trifluoride Hydrogen bromide	Hydrogen chloride Hydrogen sulfide	Sulfur tetrafluoride Phosphorus pentafluoride
Flammable	350	Carbon monoxide Ethylene	Ethane Hydrogen	Methane Natural gas
Chemically inert	580	Argon Helium	Krypton Neon	Nitrogen Xenon
Toxic	660	Cyanogen Hexafluoropropylene	Hydrogen fluoride Nitric oxide	Phosgene Sulfur dioxide
Oxidizer	540	Oxygen		

Cylinders can be equipped with excess-flow valves that shut down the gas supply in the event of abnormal flow resulting from rupture, fire, or other causes. This is critical with poisonous or flammable gases and can also be important with inert gases in small, poorly ventilated areas where asphyxiation is a potential hazard.

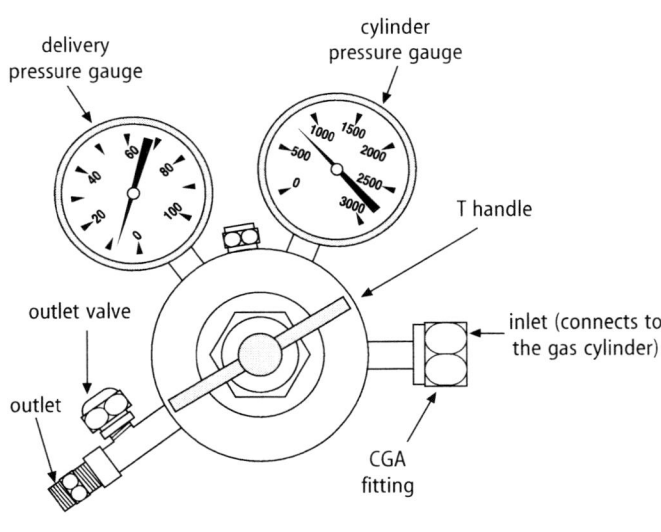

Figure 5C-2: The parts of a dual stage regulator

Pressure regulators reduce a high-pressure supplied gas to a lower pressure and maintain the flow level required for an operation. Many types of regulators are available to meet different specifications. Figure 5C-2 shows the parts of one kind of regulator. Each regulator is supplied with a specific CGA standard inlet connection to fit the cylinder valve for a particular gas. When using compressed gases, open valves slowly and only with the appropriate regulator in place and in the "off" position. Close the valve as soon as the desired amount of gas has been released. Never oil or grease regulator or cylinder valves, because these lubricating substances may be reactive with some oxidizing gases, such as oxygen and fluorine.

Check cylinders, connections, and hoses regularly for leaks. This is particularly important for flammable or toxic gases. To check for leaks you can apply soapy water (1:10 solution of dish soap and water) to the connections, hoses, and valves. Any bubbles indicate leaks.

Leaking gas cylinders may constitute serious hazards. Follow the procedures established in your workplace. Depending on the type of gas that is leaking, you may need to evacuate or immediately call for outside help.

Personal Protection for Handling Cylinders

Always wear safety goggles when handling compressed-gas cylinders. Gloves of an impermeable material that provide protection from the cold should also be worn. The gloves should be dry and free of dirt, oil, and grease and should be loose-fitting so they can be removed easily if liquids spill into them. Most companies or laboratories require employees to wear steel-toed shoes when moving cylinders; check your company or laboratory's standard operating procedure (SOP).

Transporting, Positioning, and Storing Cylinders

Transporting gas cylinders and positioning them for use requires care. Do not drag, roll, or slide cylinders or allow them to bump each other forcefully. Never move or transport cylinders with the regulator attached. Always transport them on wheeled cylinder carts with retaining straps or chains to secure the cylinders. Use these carts even for short distances. Never lift a cylinder by the cap. Position cylinders along a wall, secured firmly with a clamp and belt or chain. Leave the valve protection cap in place until cylinder is fully secured and ready for use. Cylinders of different OSHA hazard classes, such as flammables and oxygen, must be stored a minimum of 20 feet apart unless separated by a firewall or concrete-block partition.

Connecting the Regulator

When you are ready to attach the regulator, remove the cylinder cap and make sure that the valve is tightly closed. Check your regulator carefully to make sure that the CGA number on the side of the regulator matches the number for the gas you are using. For example, if you are using a corrosive gas, the regulator should be stamped with the number "330." If you aren't sure whether you have the correct regulator, ask a qualified person.

When you are connecting the regulator, never force the connection. Never use an adaptor. If the connection can't be made easily, the regulator may be the wrong one for that cylinder. **Never** use any type of lubricant on the regulator or cylinder valve. Also, **never** use Teflon® tape to seal the CGA fitting to the cylinder valve because Teflon® can burn in the presence of oxygen or other strong oxidizers.

Handling Empty Cylinders

Cylinders should never be emptied completely; slight positive pressure should remain in the cylinder to keep out contaminants, which could cause a violent reaction within the cylinder, depending on the gas it contained. According to *Prudent Practices*, a cylinder should never be emptied to a pressure lower than 25 psig or 25 psi above ambient pressure.

When a cylinder is considered empty, close the valves; remove the regulator according to standard procedures; replace all valve-protection caps, outlet dust caps, and other accessories shipped with the cylinder; clearly mark or label the cylinder as "empty"; and store it secured and in a designated area away from full cylinders to await return to the supplier.

Careless handling of empty cylinders could result in an empty cylinder being mistaken for a full one. If an empty cylinder is accidentally connected to a high-pressure system, foreign materials could be forced into the cylinder, causing a hazardous situation.

Other Cylinder Handling and Use Precautions

Some additional cylinder handling and use precautions are listed below.

- Never subject any part of a cylinder to a temperature higher than 125°F, and never allow any part of a cylinder to come in contact with a flame.

- Never subject standard cylinders to extremely low temperatures (-40°F or lower), because many types of steel lose their ductility and impact strength at such temperatures. Special stainless-steel cylinders are designed for use at extremely low temperatures.

- If cylinders are stored in the open, shield them from the ground beneath to prevent rusting. If a cylinder has become covered with ice or snow, thaw it at room temperature.

- Never place cylinders where they could become part of an electrical circuit.

- Make sure all hoses in a compressed-gas system are secured; in case of rapid release of compressed gas, an unsecured hose can whip dangerously, posing a physical hazard and possibly building up a static charge that could ignite a combustible gas.

- When working with flammable compressed gases, make sure all cylinders, lines, and equipment are bonded and grounded.

- Special precautions are required when using cylinders in earthquake zones.

A Sample Procedure for Working with Gas Cylinders

The table on the following page describes a sample procedure for working with compressed gas cylinders. These steps are based on the recommendations of the Compressed Gas Association and *Prudent Practices.* However, always follow the operating procedures required by your institution's chemical hygiene plan and all local, state, and federal regulations.

Sample Procedure for Working with Compressed Gas Cylinders
1.	Put on goggles, gloves, and protective shoe coverings. Load an empty cylinder onto the hand cart. Remember not to lift the cylinder by the cap. Fasten the cylinder in the cart with the belt or chain. Roll the cart to the designated area.
2.	Unfasten the cylinder from the cart and set it in the designated place. Secure the cylinder to a wall, bench, or other firm support with a chain or belt, or place it in a cylinder stand or rack.
3.	Remove the cylinder cap and make sure the cylinder valve is tightly closed.
4.	Remove the valve outlet cap (or cylinder plug) if present, and inspect the valve threads for damage or contamination.
5.	Check the CGA number on the cylinder valve and select the correct regulator for that CGA number. ⚠ *Regulators and valves used with oxygen must never come in contact with oil and grease.*
6.	Inspect the regulator inlet and cylinder outlet connection for foreign matter. Remove foreign matter with a clean cloth unless the cylinder contains oxygen. In the case of oxygen, open the cylinder valve slightly to blow out any dirt from the outlet connection.
7.	Close the regulator by rotating the adjusting screw or T handle in a counterclockwise direction.
8.	If the regulator is equipped with an outlet valve, close the regulator outlet valve by turning it clockwise.
9.	Attach the regulator to the cylinder valve using first your fingers and then a flat-faced or adjustable wrench.* Note that some threads are right-handed and screw in clockwise; others have to be tightened counterclockwise. ⚠ *Do not force the connection. If the connection still cannot be made easily, the regulator is probably the wrong one for that cylinder. Double-check the CGA number, and ask your instructor for assistance if necessary.*
10.	Connect tubing from the regulator outlet to the inlet of the system requiring the gas.* ⚠ *The method of connecting the regulator will vary with the kind of equipment being used. Make this connection according to the equipment instructions.*
11.	Turn on the gas supply by slowly opening the cylinder valve completely.
12.	Adjust the gas pressure to the desired pressure by turning the regulator's T handle clockwise. Open the outlet valve if the regulator is equipped with one.
13.	When finished, close the gas cylinder valve completely.
14.	Vent the regulator by opening the T handle and the outlet valve all the way. Wait a few seconds before proceeding to the next step. ⚠ *Wait until both gauges descend to zero before disconnecting the regulator from the system as described in the next step.*
15.	Disconnect the regulator from the system requiring the gas.
16.	Disconnect the regulator from the cylinder by slowly loosening the cylinder connection and listening for gas seepage. ⚠ *If you can hear gas leaking, make sure the cylinder valve is tightly closed. Also make sure the regulator has been completely drained.*
17.	Replace the plug in the cylinder valve outlet (if applicable) and replace the cylinder cap over the valve.
***Never** use copper or brass fittings with acetylene gas. These substances react to produce an insoluble, explosive compound.

Section 5D: Working with Laboratory Glassware
Instructor Notes

The objective of this section is for students to understand the factors that affect the safety of working with laboratory glassware and how to check such equipment for safety. The demonstrations "Looking at Strains in Glass" and "Prince Rupert Drops" dramatically illustrate the potential hazards of using glassware that has been stressed.

Student Pages

- Section 5D: Working with Laboratory Glassware—Background

Possible Playout of the Section

Students read the Background and discuss the safety issues associated with using laboratory glassware. You may wish to do the demonstrations to illustrate the idea that glass with no visible flaws can still pose a hazard. The "Looking at Strains in Glass" demonstration enables students to see how strain damage looks using polarized light. The "Prince Rupert Drops" demonstration shows how fragile glass can be when under stress.

Safety for Demonstrations

The demonstrations in this book do not routinely include special safety precautions for the chemicals or equipment that might be used. As the course instructor, it is expected that you will both follow SOPs and local regulations for conducting any demonstration and discuss these precautions with your class to reinforce the building of a safety culture.

Setup Notes for Demonstrations

Looking at Strains in Glass

Materials
- overhead projector
- polarizing film
   *Experimental-grade polarizing film material is available from Edmund Scientific (800/728-6999, www.edsci.com) in individual 7-inch x 7½-inch rectangles (cat. #T37,349) and 20-packs of 2½-inch-diameter circles (cat. #T38,396).*
- long glass droppers, Pasteur pipets, or other soft glass
- small item of Pyrex® glassware
- quartz glass, such as cuvets
- tongs
- Bunsen burner
- appropriate personal protective equipment (PPE)
- (optional) triangular file

Preparation

To prepare samples from soft glass and Pyrex®, carefully grasp one end with tongs, and hold the glass in the flame of a Bunsen burner. Heat the glass thoroughly and allow it to bend slightly at any point you choose. Set the glass on a heat-resistant surface to cool. The defects produced in the glass by heating will be visible under the polarizing film, but they will be subtle. You may also want to produce some more dramatic defects by scoring glass samples with a file. (Rinse glass before using.) For comparison, you may want to view quartz glass (such as cuvets) through the polarizing film. Ideally, you would not find defects in this glass.

Prince Rupert Drops

Materials
- soft glass rod
- hammer
- pliers
- paper towels
- propane torch (must have a tip that focuses flame to a sharp point)
- 1-L beaker about ¾ full of water
- appropriate personal protective equipment (PPE)

Preparation

Make the Prince Rupert Drops as follows: Hold the glass rod at about a 45° angle and position a beaker of water to catch the molten drops of glass as they fall. Rotate the glass in the hottest part of the propane torch so that as the glass starts to melt, molten drops will fall into the water. Remove the glass drops from the water, and carefully break off the thin flexible "tails" at least 1 inch from the top of the drop. (If you break the tail too close to the drop, the drop will shatter.)

Presentation of Demonstrations

Looking at Strains in Glass

Place a sample of strained glass on an overhead projector between two layers of polarizing film (see illustration), and slowly rotate one piece of the film until a strain pattern appears. The effect may be subtle, so students will need to look carefully.

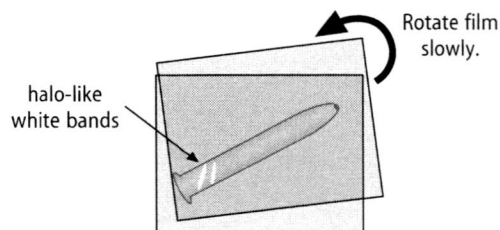

Rotate the film slowly until halo-like white bands are visible.

Prince Rupert Drops

Place a Prince Rupert Drop on an overhead projector between two sheets of polarizing film, and slowly rotate one piece of the film until a strain pattern appears. Explain that the rapid cooling of the outside of the molten drop as it falls in the water creates tremendous internal stresses that are visible through the film. To show the effect of this stress on the glass, place the drop between sheets of paper towel or a heavy cloth, and use pliers to break off the tail of the drop. The entire drop will shatter. (Shielding and/or goggles should be used during this demonstration.) Explain that although the Prince Rupert Drops break easily under tension, they do not break under compression. To demonstrate, place a drop between sheets of paper towel or a heavy cloth, and hit it firmly with a hammer. The drop will not break unless the tail is inadvertently sheared off by the blow.

Acknowledgment

Prince Rupert Drops demonstration adapted from "Demonstration: Prince Rupert Drops, Glass Under Stress," Dyer, J., *CHEM 13 News,* November 1988.

Section 5D: Working with Laboratory Glassware
Background

Note: This section provides general recommendations for and information about working with laboratory glassware. However, you should always follow the procedures established by your organization's chemical hygiene plan.

The safe use of laboratory glassware requires knowing how to inspect it for cracks or flaws before use, select and correctly assemble the glassware required for various procedures, properly clean the glassware after use, and safely clean up shards if breakage occurs.

Types of Glass

Laboratory operations involve the use of several different types of glass. Soda glass (soft) is used for common bottles and jars, borosilicate glass (hard glass, such as Pyrex® or Kimax® brands) for vessels that must withstand heating or cooling, and quartz glass (such as Vycor® brand) for spectroscopy sample containers and photolysis vessels.

Inspecting Glassware Before Use

Inspect glassware before each use for visible flaws such as cracks, star cracks, scratches, etching marks, and chips. Pay particular attention to the lips and bottoms of the glass apparatus. Laboratory glassware can also have flaws that are not visible under typical lighting conditions. These flaws (called strains) are a result of stress—forces that are capable of causing deformation (such as compression, tension, and shear) when acting on an object. Stress is expressed as force per unit area. Strain is a measure of the amount of deformation. If flaws are detected, dispose of the glassware.

Glass with strain damage can break if it is subject to torque, filled with hot liquid, used under pressure or heat, or even required to bear its own weight. Be aware that scratch marks from excessive stirring can cause glass to separate. Persons responsible for maintaining glassware use an instrument called a polariscope to check glassware regularly for strain. Glass blowers use a similar device. The patterns seen in stressed material under the polariscope show areas of strain. Glass is normally *isotropic*—that is, light travels through it at the same speed in all directions. Under strain, the internal displacement of molecules causes the glass to be *anisotropic.* In other words, the light waves move at different velocities depending on the direction. The emerging light waves are thus out-of-phase and interfere with each other to produce typical patterns.

Assembling Apparatus

Apparatus assembled from glassware components are used for many common laboratory procedures such as distillation, extraction, and refluxing. The assembly of glass apparatus has been simplified by the use of standard-taper ground-glass joints. Common sizes include 24/40 and 19/22; the first number refers to the diameter in millimeters of the larger tube and the second to the length in millimeters of the ground-glass surface. The ground-glass joints must be lubricated with a small amount of lubricant such as silicone grease. Lubrication makes the ground-glass joints easier to separate and prevents breakage. Use only a small amount of lubricant, and apply it only to the top part of the ground-glass joint.

Glass tubing or rods inserted into rubber stoppers or plastic tubing is a part of many assemblies. *Prudent Practices* states that "Cuts from forcing glass tubing into stoppers or plastic tubing are the most common kind of laboratory accident and are often serious." Using the following procedure and tips can help prevent such accidents:

- Make sure the glass tube is the proper diameter for the hose or stopper.

- Ends of tubes to be inserted should be fire polished and allowed to cool before inserting.

- Lubricate the end of the glass tube to be inserted with water or glycerol.

- Protect your hands with heavy gloves or layers of cloth.

- Hold the glass tube close to the end to be inserted (within approximately 5 cm from the end).

- Using a slight twisting motion, gently insert the glass tube as shown in Figure 5D-1. Do not use too much pressure and torque.

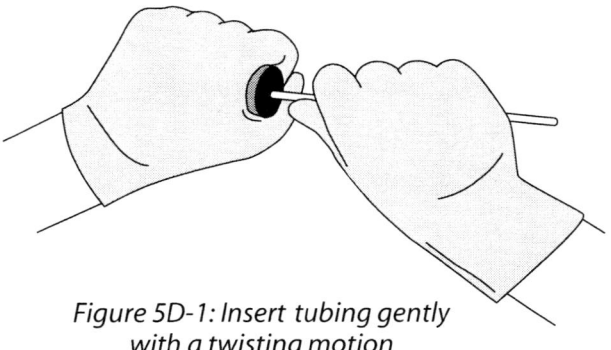

Figure 5D-1: Insert tubing gently with a twisting motion

- If a tube becomes stuck, slit the hose or stopper with a sharp knife and carefully extract the tube. Do not attempt to tug the glass forcefully.

- To further reduce risk, you may want to use a cork borer as a sleeve for the glass tubes. Also, you may want to use a commercial product such as the Glass-a-Matic Hand Saver or Safety Tips (Flinn Scientific, cat. #AP4599 and #AP4524, 800/452-1261, *www.flinnsci.com*), which are specially made to assist people with inserting glass rods into stoppers or hoses.

Other general recommendations for working with glassware safely are as follows:

- Clamp apparatus firmly and well back from the edge of the laboratory bench.

- Choose the right size of glassware for the operation, leaving at least 20% free space in the container.

- Consider the proximity of reagent bottles to burners and to other workers and their equipment.

- Never carry out reactions in or apply heat to a closed system unless it has been designed and tested to withstand pressure.

- Avoid using glassware for work at high pressure.

- If conducting pressure or vacuum operations in glass vessels, use adequate shielding to protect against violent collapse. Use only round-bottomed or thick-walled evacuated reaction vessels specifically designed for reduced-pressure work. Never evacuate thin-walled flasks, Erlenmeyer flasks, or round-bottomed flasks larger than 1 L.

- Place a pan under a reaction vessel or container to catch spilled liquids in case the glass breaks. The pan should be large enough to contain the entire spill.

Using Glassware for Reduced-Pressure Work

Reduced-pressure work poses special glassware hazards. Evacuated glass vessels can implode, resulting in flying glass and spattering chemicals. Only glassware designed for vacuum work should be used; it must be completely free from flaws and should be made from Pyrex or similar materials. Appropriate reaction vessels for low-pressure work include round-bottomed flasks (1 L or less) and thick-walled side-armed suction flasks that are wrapped with tape if possible. Glass vacuum desiccators should be made of Pyrex or similar glass and should be completely enclosed in a shield or wrapped with friction tape in a grid pattern that allows visibility of the contents. Never carry or move an evacuated desiccator unless it is wrapped with tape and placed in a specially designed holder.

When assembling vacuum apparatus, avoid strain on the joints, the necks of flasks, and the vacuum line. Support heavy apparatus from below as well as by the neck. Place the entire apparatus well back on the bench and place explosion-proof shielding around the apparatus. Goggles and face shields must also be worn by all people in the area. If the chemicals in use require ventilation, the apparatus and explosion-proof shielding should be set up in a hood and the sash lowered to the proper level (as discussed in Chapter 4). If ventilation of the chemicals is not required, a hood can also be used to provide minimal explosive protection. In such cases, the hood sash is typically closed completely. However, it is important to understand that a hood or shield does not provide adequate protection for anything more than minor explosions.

The construction of a Dewar flask involves a vacuum existing between the layers of glass. This type of construction is susceptible to breakage due to a thermal or mechanical shock.

To reduce the risk of flying glass if a collapse occurs, Dewars with exposed glass should be shielded with a layer of fiber-reinforced friction tape or housed in a wood, plastic, or metal container.

Cleaning Glassware

Cleaning glassware may expose you to flammable solvents, harsh detergents, or broken glass. Follow these precautions to reduce the hazards:

- Wear proper eye protection to protect your eyes from fumes, splashing, and glass shards resulting from breakage.

- Minimize the use of flammable solvents for cleaning. If solvents must be used, follow all precautions listed on labels, in MSDSs, and in local ordinances.

- Do not use nitric acid, sulfuric acid, or other strong oxidizers unless specifically told to do so. Explosions involving strong oxidizing solutions, such as chromic/sulfuric mixtures, have been reported. If such cleaning agents must be used, follow all precautions listed on labels, in MSDSs, and in local ordinances.

- Always wear appropriate gloves that have been visually checked to ensure that no holes are present, and wear other appropriate protective equipment as indicated by the cleaning agents you are using.

- If the bottom of the sink is not clearly visible due to turbid water, reach into the water gently, being alert for broken glassware. If you find broken glass, drain the sink and use a pair of heavy gloves to remove the broken glass. Rubber or plastic sink mats will help minimize breakage.

- When storing ground-glass joints and stoppers in the inserted position, separate the two pieces by a small strip of paper. This will prevent the joint from "freezing."

Chapter 6

Safe Handling, Storage, and Disposal of Chemicals

In a government research laboratory, a senior scientist uncovered some old bottles of ethyl ether. Having an immediate need for ether, the scientist attempted to remove the top from the can. Finding this difficult, he used a pair of steel pliers to twist the cap. Boom! He lost one eye and two fingers.

Learning by Accident, page 18

Chapter 6 Objectives

Pertinent Voluntary Industry Standards

- *VIS (L2.02):08*—Describe appropriate disposal techniques required for each of the categories of common hazardous materials.

- *VIS L3 Standards*—Describe characteristics of chemical materials that are applicable to storage and handling (toxicity, health effects, flammability, reactivity, sensitivity, and compatibility with other materials).

- *VIS L3 Standards*—Apply federal, state, and local regulations when storing and disposing of chemical materials and waste, and know where to find current information about implementing these regulations.

- *VIS L3 Standards*—Prepare materials for proper storage.

- *VIS L3 Standards*—Identify incompatible combinations of chemicals that could result in potentially dangerous situations.

- *VIS (L3.03):02*—Classify chemicals according to safety and health hazards (flammables, corrosives, oxidizers, and carcinogens).

- *VIS (L3.03):03*—Recognize and handle corrosive materials properly.

- *VIS (L3.03):04*—Use a chemical reference handbook to identify hazards associated with handling and storing chemical materials.

- *VIS (L3.03):05*—Handle and dispose of hazardous materials safely and according to regulatory guidelines.

- *VIS (L3.03):06*—Use mixing techniques appropriate for the materials, specifically when handling acids, bases, oxidizers, and strong reducing agents.

- *VIS (L3.03):07*—Store chemicals appropriately, recognizing the compatibility of the materials being stored and the containers in which they are being stored.

- *VIS (L3.03):09*—Use appropriate techniques to transfer gases, liquids, and solids from storage containers to equipment used in the laboratory.

- *VIS (L3.04):03*—Apply the special requirements for handling and disposal of radioactive materials.

- *VIS (L3.04):04*—Choose proper equipment for monitoring radioactive materials.

- *VIS (L3.05):10*—Describe potential interactions between the construction materials of a sample container and the contents being stored; identify compatible container materials for common chemicals, solutions, and mixtures.

- *VIS (L5.04):02*—Demonstrate proper techniques for pouring acids and bases and mixing them with other materials.

Objectives of the Chapter

Students will learn

- general procedures for safe handling of chemicals;

- safe handling of corrosive, flammable, highly toxic, and highly reactive or explosive chemicals;

- the importance of avoiding hazardous chemical mixtures; and

- appropriate procedures for disposal of wastes.

Why Your Students Need to Understand Safe Handling and Disposal of Chemicals

In the workplace, your students will be responsible for safely working with and disposing of hundreds, perhaps thousands, of different chemicals. The specific hazards associated with many chemicals are not known. Also, laboratory work frequently generates new substances with unknown properties and unknown toxicity. Improper handling or disposal of chemicals can result in personal injury, injury to colleagues, property damage, and violation of environmental regulations. Most industrial laboratories have detailed standard operating procedures (SOPs) that outline required handling procedures for hazardous chemicals.

However, some may not. Your students need to understand that they have a personal responsibility to know prudent practices for handling all chemicals and the special precautions to take with particularly hazardous chemicals.

Overview of the Chapter

Chapter 6: Safe Handling, Storage, and Disposal of Chemicals contains seven sections. General background information and complete instructions for the exercises are provided. The background pages are intended only to provide basic background information on the safe handling, storage, and disposal of chemicals. We recommend that you extend this background information by having students find and read the pertinent sections of the books in the safety reference library (described in Chapter 1). You may wish to have students briefly report on the information they find during the monthly safety meetings.

Section 6A: Getting to Know Safe Handling, Storage, and Disposal of Chemicals

This section introduces students to safe handling, storage, and disposal of chemicals. To initiate interest in the topic, students read and discuss Laboratory Incident Reports concerning accidents in the laboratory or workplace that resulted from improper handling, storage, and disposal of flammable and highly reactive chemicals.

Section 6B: Working with Corrosive Chemicals

The objective of this section is for students to learn safe techniques for working with acids and bases by identifying missing or inadequate safety information in written procedures.

Section 6C: Working with Flammable Chemicals

Three instructor demonstrations are provided to illustrate the importance of safe handling and storage of flammable materials. The Background provides a basic review of flammability and handling regulations for this type of chemical.

Section 6D: Working with Highly Reactive or Explosive Chemicals

This section illustrates the importance of safe handling and storage of highly reactive and explosive materials and provides a basic review of handling regulations for this type of chemical.

Section 6E: Working with Hazardous Substances

This section provides a general review of the proper handling and storage techniques used with hazardous substances. Students practice handling simulated toxic chemicals that have been treated with a fluorescent chemical marker. The instructor then uses ultraviolet light to evaluate their technique.

Section 6F: Avoiding Hazardous Chemical Combinations

The objective of this section is for students to learn how to identify reactive groups and individual chemicals that result in hazardous mixtures. Students use the free Chemical Reactivity Worksheet computer program, developed by the National Oceanic and Atmospheric Administration (NOAA) and the EPA to identify chemical combinations that will generate hazardous mixtures.

Section 6G: Categorizing Unknown Wastes for Disposal

The objective of this section is for student to learn how to characterize unknown laboratory wastes for disposal. In a paper-and-pencil exercise, students use a decision-tree flowchart from *Prudent Practices in the Laboratory* to determine the appropriate hazard category for unknown waste chemicals so that appropriate disposal procedures can be followed.

References for Chapter 6

Brennan, M.B. "Laboratory Fire Exacts Costly Toll," *Chemical and Engineering (C&E) News.* 1997, *75* (25), 29–34.

Fire Protection Guide on Hazardous Materials; National Fire Protection Association: Quincy, MA, 1997.

Hall, S.K. *Chemical Safety in the Laboratory;* Lewis: Ann Arbor, MI, 1994.

Hofstader, R.; Chapman, K. *Foundations for Excellence in the Chemical Process Industries: Voluntary Industry Standards for Chemical Process Industries Technical Workers;* American Chemical Society: Washington, DC, 1997.

Improving Safety in the Chemical Laboratory: A Practical Guide, 2nd ed.; Young, J.A., Ed.; John Wiley & Sons: New York, 1991; Chapter 18.

Kaufman, J. Laboratory Safety Institute, Natick, MA. Personal Communication, 1998.

The Laboratory Safety Institute Website: Laboratory Safety Guidelines: 40 Steps to a Safer Laboratory. www.labsafety.org/40steps.htm (accessed Feb 28, 2000).

Learning by Accident; Mojtabai, F.; Kaufman, J., Eds.; Laboratory Safety Institute: Natick, MA, 1997.

Prudent Practices in the Laboratory: Handling and Disposal of Chemicals; National Academy: Washington, DC, 1995.

Safety in Academic Chemistry Laboratories; American Chemical Society: Washington, DC, 1998.

Shakhashiri, B.Z. *Chemical Demonstrations: A Handbook for Teachers of Chemistry,* Vol. 1; University of Wisconsin: Madison, WI, 1983, pp 13–14.

Section 6A: Getting to Know Safe Handling, Storage, and Disposal of Chemicals

Instructor Notes

This section introduces students to safe handling, storage, and disposal of chemicals. To initiate interest in the topic, students read and discuss Laboratory Incident Reports concerning accidents in the laboratory or workplace that resulted from improper handling, storage, and disposal of flammable and highly reactive chemicals.

Student Pages

- Section 6A: Getting to Know Safe Handling, Storage, and Disposal of Chemicals—Laboratory Incident Reports
- Section 6A: Getting to Know Safe Handling, Storage, and Disposal of Chemicals—Background

Possible Playout of the Section

Students read the Laboratory Incident Reports and the Background and discuss the following questions:

1. What are the similarities and differences between the incidents described in the Laboratory Incident Reports?

2. What conclusions can you draw about safety from these incidents?

3. How could each of these accidents have been prevented?

Section 6A: Getting to Know Safe Handling, Storage, and Disposal of Chemicals

Laboratory Incident Reports

Incident Description: Highly Reactive and Explosive Chemicals

In a government research laboratory, a senior scientist uncovered some old bottles of ethyl ether. Having an immediate need for ether, the scientist attempted to remove the top from the can. Finding this difficult, he used a pair of steel pliers to twist the cap. Boom! He lost one eye and two fingers.

Learning by Accident, page 18

A fire resulting in $400,000 of damage occurred in 1996 at the University of Texas, Austin, when a postdoctoral fellow improperly disposed of sodium. The six-alarm fire was the 10th chemistry-related incident since 1992 involving the Austin Fire Department. The University of Texas is now spending $30 million to retrofit the building to meet contemporary fire-safety and building code standards. (See *C&E News*, Oct. 28, 1996, page 10, for more on disposal of sodium.)

C&E News, June 23, 1997

Incident Description: Flammable Chemicals

A graduate student died at Southeastern Massachusetts University when static electricity ignited her tetrahydrofuran (THF).

Jim Kaufman, Laboratory Safety Institute, personal communication

A professor and graduate student were badly injured in an explosion that occurred while pouring ether from a five-gallon can into a smaller container. Static electricity was the ignition source.

Jim Kaufman, Laboratory Safety Institute, personal communication

Incident Description: Grounding

In the lab of a petrochemical company, an employee was discharging the remaining contents of a butane cylinder in a fume hood. The fume hood itself was properly grounded, and a grounding wire had been provided within the hood. The emptying of the gas cylinder produced a static electric charge which lighted the gas, causing a flash fire and injuring the employee. The ground wire to the gas cylinder had not been used.

Jim Kaufman, Laboratory Safety Institute, personal communication

In 1989, Renaldo Hoffman placed a bucket under the valve of an ungrounded 55-gallon drum of toluene. He opened the valve and toluene fell through the air into the bucket. The free fall of the liquid created a static electric discharge explosion, destroyed the Gotham Ink Co. (Marlboro, MA), hospitalized five people, and killed 22-year-old Renaldo. Sadly, OSHA had visited the company eight years earlier and cited them for not using a ground wire.

Jim Kaufman, Laboratory Safety Institute, personal communication

Section 6A: Getting to Know Safe Handling, Storage, and Disposal of Chemicals

Background

Note: This section provides general recommendations for and information about safe handling, storage, and disposal of chemicals. However, you should always follow the procedures established by your organization's chemical hygiene plan.

In the workplace, you will be responsible for safely working with and disposing of hundreds, perhaps thousands, of different chemicals. The specific hazards associated with many chemicals are not known. Also, laboratory work frequently generates new substances with unknown properties and unknown toxicity. Improper handling or disposal of chemicals can result in personal injury, injury to colleagues, property damage, and violation of environmental and academic regulations. Industrial laboratories should have detailed standard operating procedures that outline required handling procedures for hazardous chemicals. However, they may not. You have a personal responsibility to know prudent practices for handling all chemicals and the special precautions to take with especially hazardous chemicals.

Fundamental Principles for Handling Chemicals

As outlined by *Prudent Practices in the Laboratory*, four fundamental principles underlie all prudent chemical handling practices.

- *Plan ahead:* Be sure you know the potential hazards of an experiment before beginning it. Do not begin a procedure without a plan for disposal of nonhazardous and hazardous waste.

- *Minimize exposure to chemicals:* Use personal protective equipment to prevent chemicals from contacting your skin. Use fume hoods and other ventilation devices to prevent exposure to airborne substances.

- *Do not underestimate risks:* Assume that any mixture of chemicals will be more toxic than its most toxic component. Treat all new compounds and substances of unknown toxicity as toxic substances.

- *Be prepared for accidents:* Before beginning an experiment, know what to do in case of accidental release of any hazardous substance. Know the location of safety equipment, fire alarms, and telephones. *(Emergency procedures and equipment are covered in Chapter 7.)*

Categories of Hazardous Materials

Four general categories of hazardous chemicals are listed below. A particular chemical may belong to more than one category. Knowing which category or categories a hazardous chemical belongs to will help you determine how to handle it safely.

- Corrosive chemicals cause visible destruction of or irreversible alterations at the site of contact; they are usually acids or bases. *You will learn more about handling corrosive chemicals in Section 6B: Working with Corrosive Chemicals.*

- Flammable chemicals are materials that easily ignite, sometimes spontaneously. *You will learn more about handling flammable chemicals in Section 6C: Working with Flammable Chemicals.*

- Highly reactive or explosive chemicals are chemicals that, under certain conditions, undergo rapid reactions that release energy violently. *You will learn more about handling highly reactive chemicals in Section 6D: Working with Highly Reactive or Explosive Chemicals.*

- Particularly hazardous materials include select carcinogens, reproductive toxins, and substances with a high degree of toxicity. *You will learn more about handling these materials in Section 6E: Working with Hazardous Substances.*

Hazardous Chemical Combinations

Certain classes of chemicals are known to be hazardous when combined, resulting in various combinations of explosion, fire, heat, and toxic products. Knowing these classes is an important part of planning chemical procedures and organizing chemical storage. Additionally, many individual chemicals are known to create hazards when mixed. Being familiar with lists of these chemicals can help prevent you from making a dangerous mistake. *You will practice predicting hazardous outcomes from combining chemicals in Section 6F: Avoiding Hazardous Chemical Combinations.*

Hazardous Waste Disposal Issues

Work in the chemical laboratory generates waste, some of which is classified as hazardous. A material is defined as waste when someone determines that it is no longer to be used and needs to be discarded. As a laboratory worker, you will often be the one who makes this determination and decides how the waste should be disposed of. Hazardous waste disposal is expensive and highly regulated, and it is further complicated if the chemical composition of the waste is unknown. *You will learn how to determine the appropriate hazard category for unknown waste chemicals in Section 6G: Categorizing Unknown Wastes for Disposal.*

Section 6B: Working with Corrosive Chemicals
Instructor Notes

The objective of this section is for students to learn safe techniques for working with acids and bases by identifying missing or inadequate safety information in written procedures for standardizing reagents.

Student Sections

- Section 6B: Working with Corrosive Chemicals—Background
- Section 6B: Working with Corrosive Chemicals—Exercise

Possible Playout of the Section

Students read the Background and discuss the appropriate handling of corrosive materials. Each student is assigned a different reagent for study. The students research the standard procedure(s) for making the reagent and identify any missing safety information. You may wish to give the students a specific written procedure or reference citation or instruct students to locate a reference source themselves. The students then compile a detailed list of safety protocols for preparing the reagent.

Section 6B: Working with Corrosive Chemicals

Background

Note: This section provides general recommendations for and information about safe handling and storage of corrosive chemicals. However, you should always follow the procedures established by your organization's chemical hygiene plan.

Corrosive substances cause local damage or destruction of living tissue at the site of chemical contact. Local effects can range from minor irritation to severe tissue destruction and usually occur after a single exposure. Skin and eyes are usually the sites of contact, but not always. Corrosive chemicals are often toxic and if accidentally ingested can cause damage in the mouth, esophagus, stomach, and even the intestines. Some corrosives, such as hydrofluoric acid, cause systemic damage as well as damage at the contact point.

While spills and splashes involving corrosive liquid chemicals are generally recognized as being capable of causing tissue damage, corrosive chemicals in the form of solids, gases, fumes, and vapors are also capable of causing damage. In many cases, when a corrosive solid comes into contact with living tissue, the heat generated from contact with the moisture in the tissue can be so great as to cause thermal burns in addition to the chemical destruction. If inhaled, corrosive gases or dust can damage the lining of the lungs, leading to a buildup of watery fluid (a condition known as pulmonary edema) or chemical pneumonia, and can result in disability or death. Further, since the gases and vapors given off by liquids and solids are usually invisible, they can be a hidden danger.

Classes of Corrosive Chemicals

Corrosive chemicals fall into four major classes: acids, bases, strong dehydrating agents, and strong oxidizing agents.

Acids: Concentrated solutions, aerosols, and gaseous forms of strong acids (such as HNO_3, H_2SO_4, and HCl) and weak acids (such as $HC_2H_3O_2$, $HCHO_2$, and H_3PO_4) can cause serious tissue damage to the skin, eyes, and mucous membranes. Hydrogen halide acids are serious respiratory irritants; hydrofluoric acid is especially dangerous. This acid is rapidly absorbed through the skin or by inhalation, penetrating deeply into the body tissues and causing painful, slow-healing burns and systemic damage that can lead to respiratory and heart failure. Contact with dilute solutions of hydrogen fluoride may appear painless for several hours, with serious chemical burns and bone damage showing up hours later.

Bases: Concentrated solutions, solids, and gaseous forms of metal hydroxides (such as NaOH and $CA(OH)_2$) and ammonia are corrosive and extremely destructive to both the skin and tissues of the eye. Be cautious when preparing concentrated solutions of these bases because the high heat of solution can raise the temperature to dangerous levels. The vapors of ammonia solutions are strong irritants, requiring the use of a hood. Pain from contact with bases in slow in occurring.

Strong dehydrating agents: Examples of strong dehydrating agents include P_2O_5, CaO, NaOH, and H_2SO_4. Strong dehydrating agents have a powerful affinity for water, and when mixed with water they generate a tremendous amount of heat. Proper mixing of a dehydrating agent and water requires adding the dehydrating agent to the water while stirring to prevent violent reaction and spattering. Because of their extreme affinity for water, dehydrating agents can cause serious burns to the skin.

Strong oxidizing agents: Strong oxidizing agents can also have serious corrosive effects. Examples include potassium permanganate, perchloric acid, chromic acid, peroxides, and halogens such as fluorine, chlorine, and bromine.

General Handling Techniques

In addition to using chemical splash goggles, *Chemical Safety in the Laboratory* recommends that these precautions always be observed when working with corrosive chemicals:

- Protect your hands and forearms by wearing appropriate chemical-resistant gloves, a laboratory coat, and a rubber apron to avoid any contact of corrosive chemical with any part of the body. Use chemical-resistant, disposable sleeves if a laboratory coat is not available.

- Procedures involving corrosive chemicals that may result in the generation of corrosive gases, vapors, fumes, aerosols, and dusts must be conducted in a hood or other suitable containment device. If a hood is used, it must have been evaluated previously to establish that it is providing adequate ventilation.

- Remember that even simple procedures such as dissolving pellets of NaOH in water or diluting concentrated acids can present serious safety hazards from the heat released in the process. When diluting acids, always add acid to water, never the other way around. Make additions slowly and cautiously. Setting the container in an ice bath can help control the temperature increase. Stir to guard against splattering that can result from localized heating.

- To minimize hazards from spills resulting from overturned containers or accidental breakage of apparatus, containers must be stored in pans or trays made of unbreakable, chemical-resistant material, and the apparatus must be mounted above trays of the same type of material. Trays must be large enough to contain all the chemical that could potentially spill.

- The laboratory worker must be prepared for possible accidents or spills. If a corrosive chemical contacts the skin, the area must be well irrigated with copious amounts of running water, using a safety shower if necessary. (See Chapter 7.) Some corrosives such as bromine require special precautions. Be sure to read the MSDSs and SOPs before using to ensure all necessary precautions are taken.

- All spills must be cleaned up by properly trained personnel wearing suitable personal protective equipment. The proper disposal of the cleaning materials will be determined by the nature of the spilled chemical.

- Contaminated clothing and shoes must be removed from the body. Salvaged clothes and shoes must be properly cleaned and decontaminated.

- Upon completion of the operation, workers must clean the exterior of their gloves and rubber aprons before removal.

Storage

The following storage precautions are recommended for corrosive chemicals according to *Chemical Safety in the Laboratory:*

- Prevent corrosion of metal shelves or dehydration of wooden shelves by providing acid-resistant trays under containers of corrosive chemicals.

- Emergency water and safety equipment, such as eye- and face-wash fountains and safety showers, must be available in all storage areas since many of these chemicals are corrosive to human tissues. This equipment must be tested on a regular basis.

- Acids are incompatible with bases, and oxidizing agents are incompatible with reducing agents. These groups of chemicals must be segregated. If they have to be in close proximity, they should be placed in separate cabinets in break-resistant containers.

- Supplies for the cleanup of spills and neutralizing agents appropriate for the corrosive chemicals being stored must be available near these chemicals.

Strong oxidizing agents require additional storage precautions. They must be stored in containers made of glass or other inert material, and rubber stoppers or corks should not be used. Oxidizing agents must be segregated from flammable chemicals. If they have to be in close proximity, they should be placed in separate cabinets in break-resistant containers.

Section 6B: Working with Corrosive Chemicals
Exercise

Many laboratory procedures require standardized solutions of bases or acids. These standardized solutions are made according to established methods. The objective of this Exercise is for you to analyze a published procedure for making and standardizing a reagent assigned by your instructor. Based on what you've learned about safe handling and storage of corrosive chemicals, you will identify missing safety information and conduct research to find this information.

Examine the procedure in detail to identify any safety information that is already provided. Ask yourself whether the safety information is complete. For example, the procedure might specify "chemical-resistant gloves" but not identify the actual type of gloves to use. Next, identify any missing safety instructions. Find the details on the missing safety information, and be sure to keep track of your sources. Include your organization's chemical hygiene plan among the references for your review.

Write a detailed safety protocol for the procedure you are evaluating. Refer to specific steps as well as the general overall procedure, and include the references you used.

Section 6C: Working with Flammable Chemicals
Instructor Notes

Three instructor demonstrations are provided to illustrate the importance of safe handling and storage of flammable materials. The Background provides a basic review of flammability and handling regulations for this type of chemical.

Student Pages

- Section 6C: Working with Flammable Chemicals—Background

Possible Playout of the Section

After students read the Background, you may wish to do the demonstrations in this section to dramatically illustrate three important ideas: sparks can ignite the vapor from just a few drops of alcohol (Alcohol Popper); pouring liquids can generate static electricity (Kelvin Water Dropper); and the role of ignition temperature (Nonburning Bill).

Safety

The demonstrations in this book do not routinely include special safety precautions for the chemicals or equipment that might be used. As the course instructor, it is expected that you will both follow SOPs and local regulations for conducting any demonstration and discuss these precautions with your class to reinforce the building of a safety culture.

Additionally, in the Nonburning Bill demonstration, it is important to clear the performance area of flammable and combustible materials. Also, be sure the students will be at least 10 feet away. Have a fire extinguisher ready. Use appropriate shielding.

Setup Notes for the Demonstrations

Alcohol Popper

Materials
- Piezo Popper kit (item # HS-2A, available from Educational Innovations, 888/912-7474, *www.teachersource.com*)
- methanol or other type of alcohol recommended in the kit
- electrical tape

Kelvin Water Dropper

Materials
- Kelvin water dropper
  The device can be either constructed using the drawings and directions in Scientific American *(June 1960, pp 175–185) or purchased from commercial vendors (such as Educational Innovations, 888/912-7474,* www.teachersource.com *or the Laboratory Safety Institute, 508/647-0900,* www.labsafety.org*).*

Nonburning Bill

Materials
- appropriate personal protective equipment (PPE)
- alcohol (one of the following)
  - 50 mL 99% isopropyl alcohol
  - 70 mL rubbing alcohol (70% isopropyl alcohol)
  - 50 mL 95% ethyl alcohol
- water
- 1-L beaker
- stirring rod
- dollar bill or small cloth towel
- Bunsen burner or piezoelectric gas grill starter
- tongs
- porcelain dish
- beaker of water
- (optional) table salt (sodium chloride, NaCl)

Preparation
Prepare one of the following water-alcohol mixtures:
- mix 50 mL 99% isopropyl alcohol with 50 mL water,
- mix 70 mL rubbing alcohol with 30 mL water, or
- mix 50 mL 95% ethyl alcohol with 50 mL water.

You may want to add some NaCl to the alcohol-water solution to make the flame more visible.

Presentation of the Demonstrations

Alcohol Popper

The Piezo Popper consists of a film canister filled with flammable vapor that is set up to have a spark applied when you press the starter button on a piezoelectric gas grill starter. As an introduction to this demonstration, show that alcohol can burn "safely" under proper conditions by burning it in a porcelain dish. Emphasize that the danger significantly increases when a high concentration of vapor is allowed to build up in a confined space. Follow the procedures and safety instructions provided with the Piezo Popper kit.

Kelvin Water Dropper

The Kelvin Water Dropper can be used to clearly demonstrate the presence of a commonly overlooked hazard—static electricity. The Kelvin Water Dropper is a unique device for demonstrating the generation of static electricity from the free fall of a liquid through the air. Follow the instructions and safety recommendations provided by the manufacturer of this device.

Nonburning Bill

Immerse a dollar bill in the alcohol-water mixture to wet it thoroughly. Use tongs to lift the bill out of the liquid and hold it over the beaker while the excess liquid drips off. (If using a cloth towel instead of a dollar bill, squeeze out the excess liquid and rinse your hands of any alcohol-water mixture to minimize the chance of burns.) If using a Bunsen burner, set the bill in a watch glass while you light the burner.

With the tongs, hold the dollar bill in the center at arm's length, away from yourself, others, and flammable objects. Cautiously ignite the wet bill, keeping it well away from yourself to minimize the chance of burns. You may want to ask a volunteer to lower the lights to make the flame more visible. Extinguish the flame before all the alcohol-water solution has burned off by dipping the bill in a container of water or holding it under running water. This will minimize the possibility of localized charring at the edges of the bill.

Acknowledgments

Nonburning Bill demonstration adapted from "Nonburning Towel"; *Fun with Chemistry: A Guidebook of K–12 Activities;* Sarquis, M.; Sarquis, J., Eds.; Institute for Chemical Education: Madison, WI, 1993; Vol. 1, pp 245–247.

Shakhashiri, B.Z. *Chemical Demonstrations: A Handbook for Teachers of Chemistry,* Vol. 1; University of Wisconsin: Madison, WI, 1983, pp 13–14.

Section 6C: Working with Flammable Chemicals

Background

Note: This section provides general recommendations for and information about safe handling and storage of flammable chemicals. However, you should always follow the procedures established by your organization's chemical hygiene plan.

In everyday speech, the words "combustible" and "flammable" might be used interchangeably to describe burnable (ignitable) materials. However, in the laboratory, the terms "combustible" and "flammable" have specific meanings which are discussed in the following section. Although any ignitable material must be treated with care, materials that are categorized as flammable pose increased hazards and therefore have more stringent requirements for handling and storage.

Combustible and Flammable Materials

Under the U.S. Department of Transportation (DOT) and Occupational Safety and Health Administration (OSHA), liquids that burn are categorized as either combustible (less hazardous) or flammable (more hazardous), depending on how readily they burn. Solids and gases that burn according to established criteria are categorized as flammable. There is no "combustible" category for solids and gases.

Liquids

Liquids are classified as combustible or flammable based on a measurement called **flash point or flash temperature.** Flash point is the temperature at which a liquid (or volatile solid) gives off a vapor sufficient to form an ignitable fuel-air mixture at its surface. Under OSHA definitions, liquids with a flash point at or above 100°F are called combustible, and liquids with a flash point below 100°F are called flammable. DOT's definitions of flammable liquids and combustible liquids are similar to OSHA's definitions. Examples of combustible liquids include fuel oil, kerosene, and ethyl alcohol. Examples of flammable liquids include acetone, gasoline, and toluene. Depending on flash point and boiling point, NFPA further subdivides flammable and combustible liquids into seven different classes. While flash points help establish the relative hazards of flammable liquids, we must not forget that flash points are determined using standardized procedures under controlled conditions. Under conditions outside of these standard testing methods—say, a different oxygen concentration, temperature, pressure, or ignition source—even a material without a measurable flash point may burn.

Three other terms describe the flammability of fuel-air mixtures: **lower explosive** (or flammable) **limit (LEL** or **LFL)**, **upper explosive** (or flammable) **limit (UEL** or **UFL)**, and **explosive** (flammability) **range.** These values are expressed as the percent volume of fuel in air. The LEL is the leanest fuel-air mixture (smallest percent of fuel in air) that will ignite; the UEL is the richest fuel-air mixture (largest percentage of fuel in air) that will ignite. The span between LEL and UEL is the flammability range. Usually, LEL and UEL are measured at a room temperature of 25°C. At higher room temperatures, the flammability range widens. Since flammability ranges are determined under uniform conditions that may differ from actual laboratory conditions, many laboratories limit allowable concentrations of flammable vapors in the air to an amount less than 20% of published LELs.

Solids

DOT defines a flammable solid as any solid, other than an explosive, that is liable to produce fire as a result of friction or heat retained from manufacturing or that, if ignited, burns so vigorously and persistently that it poses a serious transportation hazard. Examples include safety matches, sulfur, and metallic calcium.

Gases

DOT defines a flammable gas as "a compressed gas that satisfies the criteria for flame projection, lower flammability limit, and flammability range." *(Improving Safety in the Chemical Laboratory)* Examples include methane, propane, and hydrogen.

The Fire Tetrahedron

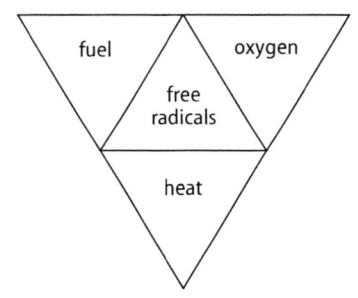

Figure 6C-1: Sample fire tetrahedron

To understand how flammable materials pose a fire hazard, you must know and understand the components required to begin and sustain a fire. Traditionally, firefighters have explained this using the concept of the fire triangle: each side of the fire triangle represents one of the components required to start a fire—fuel, oxidizer, and ignition source. To prevent fires, these components are kept apart. Today, the fire triangle idea has been expanded to include a fourth component, free radicals, the species required to sustain the fire—so that we now have what is called the fire tetrahedron. The four components are described below.

Fuels

Most common fuels are organic. These include solids such as wood, coal, paper, and plastics; liquids such as gasoline, alcohol, acetone, and many other organic solvents; and organic gases such as methane, ethane, and acetylene. Keep in mind that it is the vapors (often invisible) from flammable liquids that actually burn. Most of these vapors are denser than air and tend to flow along the ground or floor, possibly traveling more than 100 feet from the source. When vapors are ignited, the fire can travel back along the vapor trail to the source in a process called a flashback. Non-organics, including metals, are a less common type of fuel, but they cause very serious fires. The hazards of burning metals are discussed further in Section 6D: Working with Highly Reactive or Explosive Chemicals.

Oxidizers

Keeping fuels away from oxidizers requires recognizing which chemicals are oxidizers. Rudolph Gerlach, one of the authors in *Improving Safety in the Chemical Laboratory*, states that "My personal experience has allowed me to question thousands of chemical workers on the topic of oxidizers. Very few of them could name an oxidizer other than oxygen." While atmospheric oxygen at a 20% concentration is the most familiar oxidizer, several other oxidizers are listed here:

Common Oxidizers		
Solids	Liquids	Gases
nitrates	bromine	oxygen and ozone
perchlorates	hydrogen peroxide	fluorine
permanganates	nitric acid	chlorine

Ignition Sources

The ignition source provides either 1) the activation energy needed to start the chemical reaction between fuel and oxidizer, or 2) an alternative path to the product that has lower activation energy. Without the ignition source, the fuel and oxidizer could come into contact and not burn. Ignition sources include sparks from electric devices, sparks or heat from friction, open flames, lighted cigarettes or hot ashes, static electricity, hot objects, lasers, and catalysts such as platinum black that reduce the required activation energy for initiating a fire.

Substances can spontaneously ignite without a separate ignition source. The autoignition temperature of a solid, liquid, or gas is the minimum temperature required to initiate self-sustained combustion in the absence of an external ignition source (spark or flame). White phosphorus has an autoignition temperature of 30°C and can spontaneously combust at room temperature. The lower the autoignition temperature, the greater the potential for a fire. For example, laboratory professionals become quite concerned if the cooling system in their building fails on a hot summer day. They know that the room temperature could potentially reach the autoignition temperature of some of the chemicals in the stockroom. The autoignition temperature for some chemicals depends on the type of surface they are in contact with. For example, hydrazine has an autoignition temperature of 23°C when placed on a porous, rusted surface, but it has an autoignition temperature of 270°C on glass.

Free Radicals

The reaction mechanism that occurs in the burning process includes intermediate steps that produce free radicals. A free radical is a very reactive species that contains one or more unpaired electrons. Removing the free radicals as they are formed can stop the burning process. In fact, halon fire extinguishers operate by this principle. Halon fire extinguishers and fire suppression systems were widely used in the 1970s and 1980s. However, because of environmental concerns about the halogenated hydrocarbons used in them, they are no longer installed or sold. Halon fire suppression systems that were installed and are still functional are not required to be removed but cannot be recharged and must be replaced if damaged or used. (Chapter 7 discusses firefighting and the other types of fire extinguishers.)

Safe Handling of Flammable Materials

When working with flammable liquids, controlling vapor and preventing ignition are two keys to safety. Since it is the vapors of flammable liquids that burn, the concentration of vapors in the air must be kept below the LFL. Fume hoods are used to help accomplish this, especially when appreciable quantities of flammable liquids are handled in any way, such as being transferred from one container to another or being allowed to stand in open containers.

The key to preventing ignition is awareness. Chapter 5 discusses the ignition hazards posed by certain electrical devices. However, it is important to remember that not all electricity is confined to electrical appliances. Static electricity is an often unnoticed ignition source. Most people have experienced the small static electricity shocks that result from walking across carpeted floors on dry, cool days and then touching a metal doorknob. A spark occurs because of a difference in electrical potential between the hand and the doorknob. Similarly, the pouring of liquid from one container to another can result in differences in electrical potential that can discharge an electrical spark. If the liquid is flammable, the spark may contain enough energy to ignite the flammable vapors. To prevent differences in electrical potential when pouring appreciable amounts of flammable liquids (such as from storage drums into safety cans), a procedure called **bonding and grounding** is used. Bonding consists of connecting the two containers by a wire so that no difference in potential can exist between them. Since a potential can still exist between the two containers and the earth, a ground wire must be connected to one of the containers. When bonding and grounding, good metal-to-metal connections between the wires and the containers must be made. If the container is painted or rusty, the site of the connection must be cleaned before connecting the ground wire. If this is not done, the containers will not actually be bonded and grounded. Some people mistakenly think that static charge is a hazard only when the containers are metal, but this is not true. Nonmetallic containers should also be bonded and grounded in some way. When nonmetallic containers are involved in the transfer, the bonding and grounding can be made directly to the liquid rather than the container. Alternatively, some nonmetallic containers designed for flammable liquids have a small metal area designed specifically for attachment of bonding and grounding wires.

Safe Storage of Flammable Materials

Only small quantities of flammable materials should be stored in the laboratory. Specially designed safety cans for flammable liquids are recommended for holding quantities of flammable liquids greater than one liter. These cans have spring-loaded lids that prevent substantial amounts of vapor from escaping at room temperature. Safety cans should also have a flame arrestor screen that serves as a heat sink to prevent a flashback from reaching the inside of the can. The liquid level in the can should not be high enough to immerse the flame arrestor screen. The spring lid and flame arrestor should never be removed to make use of the can more convenient.

NFPA recommendations and local, state, and federal regulation prescribe what maximum quantity of flammable chemicals can be stored. Maximum container sizes and quantities that can be stored in flammable storage cabinets depend on the NFPA class of the chemical involved. Storage requirements are more stringent for chemicals with lower flash points and boiling points. For rooms where large quantities of flammable chemicals are stored and dispensed, OSHA requires mechanical ventilation, nonsparking and explosion-proof lighting and switches, and fire-resistant construction materials.

Section 6D: Working with Highly Reactive or Explosive Chemicals
Instructor Notes

The objective of this section is to illustrate the importance of safe handling and storage of highly reactive and explosive materials and provide a basic review of handling regulations for this type of chemical.

Student Pages

- Section 6D: Working with Highly Reactive or Explosive Chemicals—Background
- Section 6D: Working with Highly Reactive or Explosive Chemicals—Lab: What Ratio Gives the Loudest Pop?

Possible Playout of the Section

Two dramatic demonstrations are suggested to introduce the role particle size and surface area play on the reactivity of common chemicals in air (oxygen). The first demonstration is "Does Iron Burn?", which shows that a cast-iron frying pan and a standard iron nail do not burn in a standard Bunsen flame, while steel wool and iron filings do. Another demonstration that drives home the danger associated with dust explosions in coal mines, flour mills, and grain elevators is found in Shakhashiri's *Chemical Demonstrations* Vol. 1, pp 103–105. The Shakhashiri demonstration uses Lycopodium powder, available in magic shops as dragon's breath. However, many people are allergic to Lycopodium, and precautions must be taken to minimize their exposure to it. Both of these demonstrations require the use of appropriate protection, including shielding.

Students read the Background and complete the Lab "What Ratio Gives the Loudest Pop?", which involves micro-scale explosions in pipet bulbs filled with mixtures of hydrogen and oxygen. Students should learn that the explosive potential is greatest when the flammable gas and oxygen are mixed at a certain ratio.

Safety for the Demonstration

The demonstrations in this book do not routinely include special safety precautions for the chemicals or equipment that might be used. As the course instructor, it is expected that you will both follow SOPs and local regulations for conducting any demonstration and discuss these precautions with your class to reinforce the building of a safety culture.

Safety for the Lab

The laboratory exercises in this book do not routinely include special safety precautions for the chemicals or equipment that might be used. As the course instructor, it is expected that you will provide students with access to SOPs, MSDSs, and other resources they need to safely work in the lab while meeting all regulatory requirements. Before doing any of the laboratory exercises from this book or from other sources, we recommend that you regularly review special handling issues with students, allow time for questions, and then assess student understanding of these issues.

Setup Notes for the Demonstration "Does Iron Burn?"

Materials
- appropriate personal protective equipment (PPE), including shielding
- Bunsen burner
- iron frying pan
- iron nail
- fine-gauge steel wire
- steel wool
- iron filings
- oven mitt
- tongs

Presentation for the Demonstration "Does Iron Burn?"

With a Bunsen burner, heat first an iron frying pan and then an iron nail. Discuss differences observed. Next heat steel wire and then steel wool. (The steel wire and steel wool may burn.) Emphasize the effect of surface area on the reactivity.

Finally, sprinkle no more than a pea-sized amount of iron filings over the flame. The flame will flare up as the fine powder is lifted by the hot air and catches fire with sparking and popping. Point out that the change in behavior in the same material (from the frying pan to the iron filings) was due to increased surface area/mass ratio. Explain that most metals, when dispersed as dust, can burn with explosive force if an ignition source is present. Point out that other finely divided materials (such as dried flour or sawdust) can also burn explosively and that the potential for explosions in grain silos is a serious hazard.

Setup Notes for the Lab "What Ratio Gives the Loudest Pop?"

Materials per student
- appropriate personal protective equipment (PPE)

Materials per group
- incremented gas collection bulb, made from a plastic pipet
  Pipets are available from Micro Mole Scientific, 509/545-4904, www.micromole.com *(#P-717).*

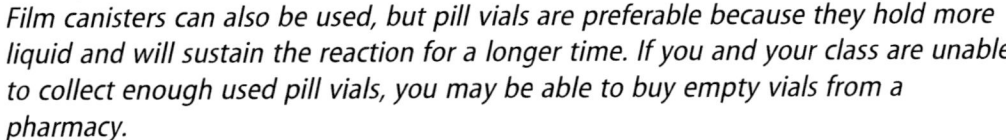

- spark igniters, made from the following
  - 10-cm length of 24 AWG speaker hookup wire (solid, not stranded)
    This wire is available from Radio Shack (#278-1509).
  - 35-mm film-canister lid
  - hot-melt glue
  - electrical tape
- micro O_2 and H_2 generators (with nozzle caps), made from the following
  - 2 plastic pill vials with caps
    Film canisters can also be used, but pill vials are preferable because they hold more liquid and will sustain the reaction for a longer time. If you and your class are unable to collect enough used pill vials, you may be able to buy empty vials from a pharmacy.
  - several grams of manganese metal (Mn)
  - several grams of zinc metal (Zn)
  - 2 plastic pipet bulbs
- bottle of 3% H_2O_2
- bottle of 1 M HCl
- Petri dish bottom
- piezoelectric gas lighter such as the Scripto Aim 'n' Flame™ Torch lighter
  Aim 'n' Flame lighters are available in grocery and discount department stores. The lighters should be empty, with no residual butane left in them.

Materials for setup
- permanent marker
- scissors
- wire stripper
- 1 of the following
  - drill and ⅛-inch drill bit
  - 6D nail, candle, or Bunsen burner, and oven mitt
- masking tape and pen for labels
- hot-melt glue gun and glue
- electrical tape

The spark igniters and micro O_2 and H_2 generators (with nozzle caps) that you prepare for "What Ratio Gives the Loudest Pop?" can be saved for future use.

Prepare the gas collection bulb as follows:
Use a permanent marker to mark each bulb in six equal parts, as shown in the illustration.

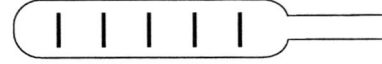

Mark the pipet bulb.

Prepare "nozzle caps" for the pill vials as follows:

Drill a hole in the center of each pill vial cap (or heat a 6D nail and push the hot nail through the center of the cap). Cut about 2.5 cm of the tip off the ends of the two pipet bulbs. Set aside the bulb ends. Push each tip end through the hole in a pill vial lid so that approximately 2 cm of the bulb tip is above the lid, as shown in the illustration.

The bulb tip should fit snugly in the hole. It may help to twist the bulb tip as you push it through the lid.

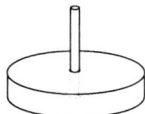

Prepare the nozzle cap.

Prepare the O_2 and H_2 generators as follows:

Prepare the O_2 generator by placing several pieces of Mn metal in a pill vial and placing a nozzle cap on the vial. Label this vial "O_2 generator." Prepare the H_2 generator by placing several pieces of Zn metal in a second pill vial and placing a nozzle cap on the vial. Label this vial "H_2 generator."

Prepare the spark igniters as follows:

Using scissors, split the end of the speaker wire down the center for a distance of 3.5 cm. Strip the last 2.3 cm of insulation from both sides of this "fork." Then strip about 0.5 cm of insulation from both sides of the other end, as shown in the illustration.

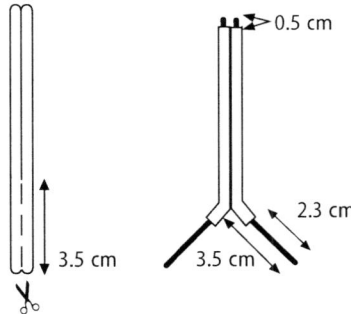

Prepare the spark igniter.

Drill a ⅛-inch hole through the center of the film-canister lid (or melt a hole with a hot 6D common nail). Slide the wire through, and secure the lid in place about 4 cm from the un-forked end of the wire with some hot-melt glue or other waterproof adhesive on the underside of the lid. (The film-canister lid serves as both a support pad and a splash guard.) (See the following illustration.)

Slide one of the two stripped wire ends down inside the small butane nozzle of the piezoelectric lighter. (See the illustration on the next page.) It should be inserted all the way down so that the insulation reaches the nozzle. (It helps to rotate the wire as you insert it.) Lay the other stripped end alongside the metal barrel of the lighter and secure it in place with three to four wrappings of electrical tape.

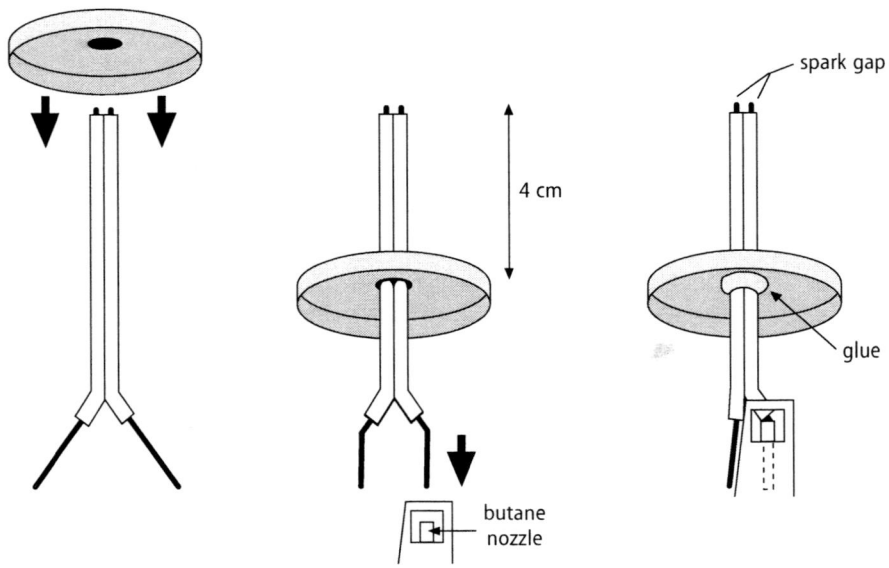

Assemble the support.

Test the spark igniter by filling a bulb with oxygen and hydrogen as described in the Student Instructions, placing the bulb over the wire so the tip of the bulb touches the film-canister lid, holding the bulb in place with your fingers, and pulling the trigger. The gas inside the bulb should ignite and explode with a pop. Be sure to hold onto the bulb when you pull the trigger or the bulb will fly across the room.

Acknowledgments

"Does Iron Burn?" demonstration adapted from "Surface Area and Burning"; *Fun with Chemistry: A Guidebook of K–12 Activities;* Sarquis, M.; Sarquis, J., Eds.; Institute for Chemical Education: Madison, WI, 1993; Vol. 2, pp 211-215.

Robert Becker, Kirkwood High School, St. Louis, MO

Lynn Hogue, Center for Chemical Education, Miami University Middletown, Middletown, OH

Section 6D: Working with Highly Reactive or Explosive Chemicals

Background

Note: This section provides general recommendations for and information about safe handling and storage of highly reactive or explosive chemicals. However, you should always follow the procedures established by your organization's chemical hygiene plan.

Under certain conditions, highly reactive or explosive chemicals undergo rapid reactions that produce a violent release of energy. These reactions can be spontaneous or initiated by certain conditions of light, mechanical shock, flames, sparks, heat, and certain catalysts, and they can produce hazardous pressures, gases, and fumes.

The OSHA Hazard Communication Standard defines a reactive or unstable chemical as a substance that "in the pure state, or as produced or transported, will vigorously polymerize, decompose, or condense, and will become self-reactive under certain conditions of shocks, pressure, or temperature." A chemical is considered water-reactive if it "reacts with water to release a gas that is either flammable or presents a health hazard."

The OSHA Laboratory Standard defines an explosive chemical as a substance "that causes a sudden, almost instantaneous release of pressure, gas, and heat when subjected to sudden shock, pressure, or high temperature."

Types of Highly Reactive or Explosive Chemicals

Reactivity is defined as the ability of a chemical to undergo a chemical reaction. Reactive chemicals can be divided into two basic categories based on their chemical structure: 1) those that exhibit instability in the absence of contact with appreciable amounts of other chemicals and 2) those that exhibit a high level of reactivity in combination with other chemicals or materials, including air and moisture.

Organic peroxides are examples of chemicals in category 1 and are particularly unstable and, therefore, especially hazardous to work with. They are actually low-power explosives and are extremely sensitive to shock, sparks, heat, impact, friction, light, and strong oxidizing and reducing agents. All organic peroxides are extremely flammable. They are frequently formed due to the instability of certain chemicals, called peroxidizable chemicals.

Examples of chemicals in category 2 include those that react on contact with water. Lithium and finely divided magnesium burn so intensely that if they are burning and come into contact with water, they will separate the water into hydrogen and oxygen molecules. The hydrogen burns and the oxygen supports combustion, so adding water to this kind of fire intensifies the fire. Sodium, potassium, cesium, and some other chemicals react violently with water to produce highly flammable hydrogen gas.

The following are other examples of highly reactive or explosive behavior in chemicals:

- Some chemicals decompose when heated. For example, certain peroxides decompose almost instantaneously when heated. Slow decomposition of a chemical may not be noticeable. If decomposition occurs on a large scale or in a confined area, the heat and gases that evolve can lead to an explosion.

- Hydrogen and chlorine react explosively in the presence of ultraviolet light.

- Many substances catalyze the explosive polymerization of acrolein.

- Shock-sensitive materials include acetylides, azides, organic nitrates, nitro compounds, perchlorates, and many peroxides.

- Many metal ions can catalyze the violent decomposition of hydrogen peroxide.

General Handling Techniques

The following are general techniques for handling highly reactive or explosive chemicals:

- Do a risk assessment and use the protective equipment required.

- Keep emergency equipment at hand.

- Assemble your apparatus so that if a reaction runs away, it is possible to remove heat, cool the system, stop adding reagents, and close the fume hood.

- Use an explosion shield and hood window in addition to goggles and a face shield.

- Do not use excessive concentrations or amounts of reagents.

- Carefully control the rate of reagent addition.

- In general, the rate of a reaction increases as the temperature increases. In exothermic reactions, increased rate will generate still more heat, possibly leading to a "runaway" reaction. Therefore, the heat evolved in a reaction must be dissipated safely.

- Keep the temperature under control with sufficient cooling and surface for heat exchange; this is important to consider when scaling up experiments in the laboratory.

- Know that some chemicals decompose when heated, and an explosive situation may develop.

- Know the violent behavior of oxidizing agents.

- Perchloric acid is a powerful oxidizing agent that may react explosively with organic compounds and other reducing agents. It must be used only in a water wash-down fume hood of noncombustible construction. Frequent inspections should be made to prevent perchloric acid and perchlorate accumulation in the exhaust system of the fume hood. Do not use perchloric acid near wooden tables or benches. Perchloric acid should not be kept in inventory; rather, it should be ordered on an as-needed basis.

- Dry picric acid is highly explosive and should be purchased only when specifically required and with a thorough understanding of its hazards. Although moist picric acid and picric acid solutions are not explosive, water may evaporate from them to leave a hazardous solid. To help avoid this problem, picric acid should be purchased in small quantities on an as-needed basis and kept on a maintenance schedule to assure it does not dry out. A proactive approach that includes dating all containers and providing a "dispose after" date can reduce the potential for explosions. Old containers of picric acid, occasionally found in stockrooms, should be disposed of only with expert assistance.

Techniques for Handling Peroxidizable Chemicals

The following are general guidelines for proper handling of peroxidizable chemicals as outlined in *Chemical Safety in the Laboratory:*

- All peroxidizable chemicals must be prevented from peroxide formation to the extent possible. Preventive measures include storage in opaque containers and the exclusion of air, preferably by a nitrogen atmosphere.

- All peroxidizable chemicals must be tested for peroxide presence before any heating or evaporation is attempted.

- The sensitivity of most peroxides to shock and heat can be reduced by dilution with inert solvents, such as aliphatic hydrocarbons.

- Metal spatulas must not be used to handle peroxides because contamination by metals can lead to explosive decomposition. Ceramic or wooden spatulas may be used.

- Friction, grinding, and all forms of impact must be avoided near peroxides, especially solid peroxides. Glass containers with screw-cap lids or glass stoppers are not to be used. Polyethylene bottles with screw-cap lids may be used.

- To minimize the rate of decomposition, peroxides must be stored at the lowest possible temperature consistent with their solubility or freezing point. Liquid or solutions of peroxides are not to be stored at temperatures lower than the temperature at which they freeze or precipitate because peroxides in these forms are extremely sensitive to shock and heat. If peroxides are stored in a refrigerator, the refrigerator must be of explosion-proof design.

- Pure peroxides must never be disposed of directly. Peroxides must be diluted before disposal.

Techniques for Handling Water-Reactive Chemicals

The following are general guidelines for proper handling of water-reactive chemicals:

- Take extreme measures to prevent accidental contact with water.

- Eliminate all sources of water in storage areas, including automatic sprinkler systems.

- Store water-reactive chemicals in a fire-resistant facility away from other combustible materials.

Section 6D: Working with Highly Reactive or Explosive Chemicals

Lab: What Ratio Gives the Loudest Pop?

Flammable gases like H_2 can ignite and burn rapidly in the presence of O_2. The quick, exothermic nature of this combustion often results in an explosion. In this lab, you generate oxygen and hydrogen and observe the loudness of the explosion generated by their reaction. You test the gases both separately and in mixtures of varying proportions.

Safety

In a laboratory setting, you are ultimately responsible for your own safety and for the safety of those around you. In order to more closely resemble an actual workplace, specific safety information is not routinely provided to you for the labs in this book. As discussed throughout the book, many resources are available to assist you in obtaining the information you need. It is your responsibility to specifically follow your institution's standard operating procedures (SOPs) and all local, state, and national guidelines on safe handling and storage of all chemicals and equipment you may use in the labs. This includes determining and using the appropriate personal protective equipment (e.g., goggles, gloves, apron). If you are at any time unsure about an SOP or other regulation, check with your instructor.

Procedure

1. Put on the appropriate PPE. Fill the bottom of a Petri dish three-quarters full of water. This will serve as a water supply during the experiment.

2. Remove the nozzle cap of the O_2 generator and add enough 3% H_2O_2 to fill the vial to within 1 cm of the top. (The MnO_2 coating that forms on the surface of the Mn metal in the vial is a catalyst for the decomposition of H_2O_2.) Replace the cap and set the generator in the Petri dish of water.

3. Remove the nozzle cap of the H_2 generator and add enough 1 M HCl to fill the vial to within 1 cm of the top. (The Zn metal in the vial reacts with the HCl to generate the H_2.) Replace the cap and set the H_2 generator in the Petri dish alongside the O_2 generator.

4. Draw water up into the incremented collection bulb to completely fill it with water. If any air is present, invert the bulb tip-upward, squeeze out any remaining air, and hold it in this squeezed position while you lower the tip under water, then release to draw up additional water.

5. Place the completely filled collection bulb tip-downward over the nozzle of the O_2 generator. (See Figure 6D-1.) Fill the bulb completely with oxygen (i.e., until no water remains in the bulb).

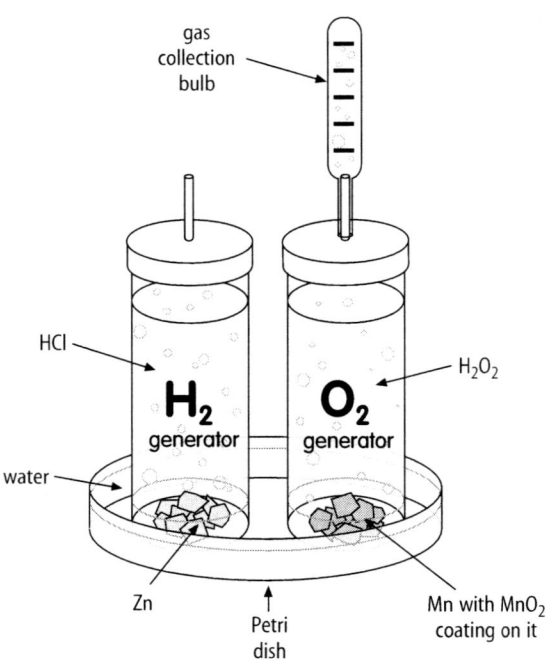

Figure 6D-1: Collect gas with your pipet.

6. Place the gas-filled collection bulb over the spark igniter as shown in Figure 6D-2. Hold onto the bulb and then pull the trigger. Rate the loudness of the pop using a scale of 0–10 to indicate relative loudness.
    Don't let go of the bulb when you pull the trigger, or the bulb may fly across the room.

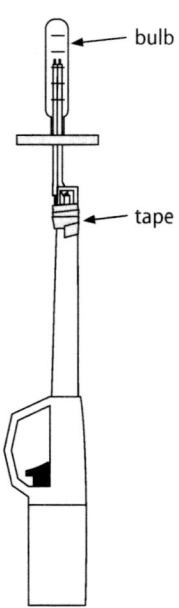

Figure 6D-2: Put the collection bulb on your spark igniter.

7. Refill the pipet bulb completely with water. Fill the bulb completely with H_2 all the way to the bottom of the tip (no water). Test and rank the loudness as before.

8. Repeat the activity for the oxygen and hydrogen ratios indicated on the Data Recording Table below.

    The pipet bulb fills with gas very rapidly. Be ready to remove the bulb from the nozzle quickly when the gas reaches the desired mark. If you accidentally collect too much of one of the gases, empty the bulb, fill it with water, and try again.

9. When you have determined the ratio that produces the loudest pop, fill your pipet bulb with that mixture. Place the bulb on the spark igniter and **aim the bulb away from bystanders**. Let go of the bulb, pull the trigger, and watch your "rocket" fly!

10. Dispose of the H_2O_2 and HCl according to your instructor's directions, leaving the metals inside the vials. Rinse the metals in the vials with water, decant off the water, replace the caps on the vials, and return as directed by your instructor.

Data Recording Table: Gas Ratios and Relative Loudness of Explosions (scale of 1–10)							
Parts O_2	6	5	4	3	2	1	0
Parts H_2	0	1	2	3	4	5	6
Relative loudness	1	4	6	8	10	6	3

Questions

1. Write a balanced equation for the reaction taking place inside the O_2 generator.
    $2H_2O_2 \longrightarrow 2H_2O + O_2 \ (g)$

2. Write a balanced equation for the reaction taking place inside the H_2 generator.
    $2HCl + Zn \longrightarrow ZnCl_2 + H_2 \ (g)$

3. Which do you think will have to be replaced first, the Mn in the O_2 generator or the Zn in the H_2 generator? Explain your reasoning.
    The Zn will have to be replaced first, because the Mn is a catalyst and does not get used up in the reaction.

4. Explain your observations for the loudness of the pop test of pure O_2.
    No pop was heard because O_2 itself is not flammable.

5. Explain your observations for the loudness of the pop test of pure H_2.
    Only a small pop was heard because H_2 gas, while flammable, requires the presence of an oxidizer to combust.

6. What is the source of oxygen that is responsible for the pop when pure H_2 is tested?
    The source of the oxygen is the air.

Section 6D: Working with Highly Reactive or Explosive Chemicals
Lab: What Ratio Gives the Loudest Pop?

200

7. What proportion of H_2 to O_2 produced the most explosive mixture? Why was that mixture the most explosive?

The loudest pop occurred when the $H_2 : O_2$ ratio was 2:1. This is the stoichiometric ratio for the reaction.

8. Write a balanced equation for the reaction between O_2 and H_2.

$2H_2 (g) + O_2 (g) \longrightarrow 2H_2O (g)$

9. What does this exercise illustrate about safe handling of gases with the potential to burn explosively?

Sparks can ignite a flammable gas if oxygen is present. An explosion can result if the reaction mixture with O_2 is within the explosive limits.

Section 6E: Working with Hazardous Substances
Instructor Notes

The Background provides a general review of the proper handling and storage techniques used with hazardous substances. Students practice handling simulated toxic chemicals that have been treated with a fluorescent chemical marker. The instructor then uses ultraviolet light to evaluate their technique.

Student Pages

- Section 6E: Working with Hazardous Substances—Background
- Section 6E: Working with Hazardous Substances—Lab: Special Handling Required

Possible Playout of the Section

Students read the Background and review the different categories of toxic substances. Before starting the Lab, emphasize to students that although the chemicals used in the Lab pose little hazard, the Lab is intended to provide experience with procedures that would be used to handle highly toxic substances. Explain that the paper face masks are being used to represent the respirators that are often required when working with highly toxic materials. Point out that the students are not wearing real respirators during the Lab because a respirator should never be worn unless a proper fit-test has been done and appropriate training completed.

The Lab is written so that after step 4 and again at the end of the Lab, you can inspect the student and the work areas with ultraviolet (UV) radiation for the presence of contamination. You can use your findings as a starting point for a discussion of ways to improve techniques and lessen the frequency of contamination. You can also discuss the proper use of UV lamps and dangers associated with their use.

Safety

The laboratory exercises in this book do not routinely include special safety precautions for the chemicals or equipment that might be used. As the course instructor, it is expected that you will provide students with access to SOPs, MSDSs, and other resources they need to safely work in the lab while meeting all regulatory requirements. Before doing any of the laboratory exercises from this book or from other sources, we recommend that you regularly review special handling issues with students, allow time for questions, and then assess student understanding of these issues.

Setup Notes

Per student
- appropriate personal protective equipment (PPE)
- small spatula or scoop
- 4-dram vial containing 2 g "Compound A" (see below)
- empty vial with tight-fitting cap
- bottle of acetone
- 20 g sucrose
- 2, 125-mL Erlenmeyer flasks with stoppers
- 2, 100-mL bottles with tops
- 100-mL graduated cylinder
- pipet
- large test tube
- water
- 3 tares—weighing papers or boats
- 400-mL beaker or container labeled "Toxic Waste"
- masking tape and pen for labels

Materials per class
- UV lamp
- balance
- "Compound A," made from the following
  - 50 g powdered sugar
  - 1 g Fluorescent Brightener 28 (enough for a class of 25 students)
    Available from Sigma Chemical Company, product number F3397, 800/325-3010, http://www.sigmaaldrich.com/.
- beakers
- red food color
- stirring rod

Prepare "Compound A" by mixing 1 g Fluorescent Brightener 28 with 50 g powdered sugar in a beaker until homogenous. Distribute about 2 g of "Compound A" into one appropriately labeled 4-dram vial for each student.

Section 6E: Working with Hazardous Substances

Background

Note: This section provides general recommendations for and information about safe handling and storage of hazardous substances. However, you should always follow the procedures established by your organization's chemical hygiene plan.

A toxic chemical is any chemical that causes adverse effects on living organisms. As discussed in Chapter 3, all toxicological considerations are based on the dose-response relationship. Some chemicals pose a particularly great risk of undesired effects on human health. Under OSHA's Laboratory Standard, additional protection must be provided for employees working with these particularly hazardous materials, including select carcinogens (an OSHA-defined category), reproductive toxins, and substances with a high degree of acute toxicity.

Select Carcinogens

Carcinogens are substances capable of causing cancer—the uncontrolled cell growth that can occur in any organ. Carcinogens tend to have a chronic (long-term) effect. The current thinking is that carcinogens interact with or modify cell DNA, changing patterns of cell growth.

Substances known to pose the greatest carcinogenic hazard are called select carcinogens under OSHA. To be classified as a select carcinogen, a chemical must meet at least one of several criteria listed in the Laboratory Standard. Examples of select carcinogens include acrylamide, asbestos, benzene, chromium(VI), formaldehyde, mustard gas, and vinyl chloride.

Although some substances are known carcinogens, many chemicals encountered in the laboratory have not been tested for carcinogenicity. Laboratory workers must evaluate the possibility that the chemical in question is carcinogenic. Sometimes, a reasonable judgement can be made by knowing the specific classes of compounds and functional group types that have been associated with carcinogenic activity. If an untested material falls into the same class as a known carcinogen, it may be considered a suspected carcinogen.

Reproductive and Developmental Toxins

Reproductive toxins have a harmful effect on various aspects of reproduction and can affect fertility and general reproductive performance. Many reproductive toxins have a chronic effect, causing damage after long, repeated exposures. The effects may become evident only after long latency periods. Developmental toxins act during pregnancy to cause death, malformations, retarded growth in the fetus, or postnatal functional deficiencies. Developmental toxins often have the greatest impact during the first trimester of pregnancy, when a woman may be unaware that she is pregnant. Women who are pregnant or may become pregnant must seek expert advice before conducting laboratory work.

Chemicals with a High Degree of Acute Toxicity

Acute toxicity is the ability of a chemical to cause immediate injury after only one exposure. As discussed in Chapter 3, LD_{50} and LC_{50} values are measures that indicate the risk of acute toxicity. The table below shows LD_{50} and LC_{50} values that fit the OSHA definition of "highly toxic" substances. Examples include chlorine, hydrogen cyanide, nitrogen dioxide, ozone, sodium cyanide, and other cyanide salts.

Acute Toxicity Hazard Level for Highly Toxic Substances			
Oral LD_{50} (rats, per kg)	Skin Contact LD_{50} (rabbits, per kg)	Inhalation LC_{50} (gas or vapor) (rats, ppm for 1 hour)	Inhalation LC_{50} (mist, fume, or dust) (rats, mg/L for 1 hour)
<50 mg	<200 mg	<200	<2
Source: 29 CFR 1910.1200, Appendix A			

General Handling Techniques

Prior to beginning work with any particularly hazardous substance (as defined by OSHA), consult a good-quality MSDS or other reliable source for toxic properties and proper handling procedures. In the workplace, your laboratory should have explicit detailed procedures for handling these types of substances. Some general guidelines from *Chemical Safety in the Laboratory* are outlined here:

- All containers of highly toxic chemicals must have labels that identify the contents and include a warning to describe the type of toxicity if known, such as "cancer-causing agent."

- All experiments with and transfers of highly toxic chemicals or mixtures must be performed in an area designated for the use of such chemicals or mixtures. Designated areas must be clearly marked with a conspicuous sign, such as "warning: toxic chemicals in use."

- Laboratory workers handling toxic chemicals must wear proper gloves and other PPE to avoid any skin contact.

- Surfaces or bench tops on which highly toxic chemicals are handled must be protected from contamination by the use of chemically resistant trays or pans that can be decontaminated after the operation. Alternatively, disposable absorbent paper liners may be used.

- If disposable apparel or paper liners have been used, these items must be placed in the proper waste container. Nondisposable protective apparel must be thoroughly washed, and containers of disposable items must be incinerated.

- By the end of the day and before leaving the laboratory, all surfaces and bench tops on which highly toxic chemicals have been handled must be properly decontaminated.

- If a spill occurs, it must be cleaned up and the area decontaminated immediately.

- Collect wastes and other contaminated materials from all experiments involving highly toxic chemicals together with the washings from flasks and other containers used. All laboratory waste must be placed in closed, suitably labeled containers for proper disposition.

Storage

Storage areas for hazardous substances must have limited access, and special signs must be posted. These storage areas must be maintained at negative pressure with respect to surrounding areas in order to prevent any airborne materials from escaping the storage area. Exhaust air from these areas must not be returned to the air supply of the building.

Section 6E: Working with Hazardous Substances
Lab: Special Handling Required

The objective of this Lab is to practice chemical handling techniques for working with toxic chemicals. In this Lab you prepare solutions of "toxic" Compound A. Be sure to follow proper precautions to minimize your and others' exposure to this material.

Safety

In a laboratory setting, you are ultimately responsible for your own safety and for the safety of those around you. In order to more closely resemble an actual workplace, specific safety information is not routinely provided to you for the labs in this book. As discussed throughout the book, many resources are available to assist you in obtaining the information you need. It is your responsibility to specifically follow your institution's standard operating procedures (SOPs) and all local, state, and national guidelines on safe handling and storage of all chemicals and equipment you may use in the labs. This includes determining and using the appropriate personal protective equipment (e.g., goggles, gloves, apron). If you are at any time unsure about an SOP or other regulation, check with your instructor.

Procedure

1. Put on the appropriate PPE. Carefully weigh 500 mg ± 5 mg of Compound A into an empty vial. Cap and appropriately label the vial.

2. Prepare stock solution A1 as follows: In an Erlenmeyer flask, prepare 80 mL of a 6% (mass/vol) solution of Compound A (from the vial in step 1) in acetone. Add two drops of red food color, stopper, and shake to mix. Transfer to a 100-mL storage bottle and label appropriately.

3. Prepare stock solution A2 as follows: In an Erlenmeyer flask, prepare 80 mL of a 20% (mass/vol) aqueous sucrose solution. Add the rest of Compound A that remains in the vial from step 2. Stopper the Erlenmeyer flask and shake to dissolve the Compound A. Transfer the resulting solution to a 100-mL storage bottle and label appropriately.

4. Pipet 10 mL solution A2 into a large test tube. Stand the test tube in a beaker so it won't spill. Cautiously pipet 6 mL solution A1 on top of the A2 layer. Stopper. Leave for your instructor to observe.

5. Leave the storage bottles for solutions A1 and A2 in your area. Clean up the area, but do not return any equipment or glassware you used to its proper place until instructed to do so.

Section 6F: Avoiding Hazardous Chemical Combinations
Instructor Notes

The objective of this section is for students to learn how to identify reactive groups and individual chemicals that result in hazardous mixtures. Students use the free Chemical Reactivity Worksheet computer program (developed by NOAA and the EPA) to identify chemical combinations that will generate hazardous mixtures.

Student Pages

- Section 6F: Avoiding Hazardous Chemical Combinations—Background
- Section 6F: Avoiding Hazardous Chemical Combinations—Exercise

Possible Playout of the Section

For the Exercise, students use the Chemical Reactivity Worksheet software, developed by the Computer-Aided Management of Emergency Operations (CAMEO) Team at the Hazardous Materials Response and Assessment Division, the National Oceanic and Atmospheric Administration (NOAA), and the Chemical Emergency Prevention and Preparedness Office of the EPA. The Exercise explains how to download and use the software. Direct comments or questions about the Worksheet to *ReactivityMail@hazmat.noaa.gov* or to the Hazardous Materials Response and Assessment Division, NOAA, 7600 Sand Point Way NE, Seattle, WA 98115, 206/526-6317.

Note that the Worksheet software is very sensitive to spelling. Unless the chemical name is typed exactly as listed in the Exercise, the Worksheet may fail to identify the chemical. The Worksheet also tends to retrieve all chemical names that have the same word within them. For example, a query for "phosphorus" will not only retrieve various forms of elemental phosphorus, but many phosphorus compounds as well. The students should be aware of this feature.

The Worksheet provides a good opportunity to review the use of CAS Registry Numbers in chemical identification. Students learn the various synonyms that exist for a given chemical product and the value of having a unique identification number. You may wish to have the students enter the chemicals into the Worksheet by CAS Registry Number to show how much faster it is to identify the specific chemical by this method.

Section 6F: Avoiding Hazardous Chemical Combinations

Background

Note: This section provides general recommendations for and information about avoiding hazardous chemical combinations. However, you should always follow the procedures established by your organization's chemical hygiene plan.

Some chemical combinations can result in the generation of large amounts of heat, violent explosions, or the evolution of flammable or toxic gases. Safe laboratory practices minimize the chance that incompatible chemical groups will accidentally mix. Even though incompatible chemicals may seem to be safely separated by their individual containers, incidents like inadvertent jarring; fire; explosion; or building vibrations from construction, earthquakes, high winds, or slamming of doors can result in a hazardous mixture. Thus, keeping incompatible chemicals segregated is a key safety practice.

Since chemical storerooms and stockrooms are places where a wide variety of chemicals are found, storage must be carefully planned. While alphabetical storage of chemicals may seem like a simple solution, it all too often results in incompatible chemicals being stored near each other. For example, cupric sulfide and cadmium chlorate explode on contact, and stannic chloride and turpentine will undergo a flame-producing, exothermic reaction.

Poor laboratory housekeeping, such as accumulation of dispensing bottles in hoods or on benchtops, can also lead to close proximity between incompatible chemicals.

Chemically Related and Compatible Groups

Planning to prevent hazardous combinations requires knowing which groups of chemicals are compatible. Determining the general class (acids, bases, flammables, oxidizers, etc.) to which each chemical belongs is a first step. However, not all chemicals within a broad group are compatible. For example, some acids are oxidizing agents (such as nitric acid) and others are reducing agents (such as glacial acetic acid); and some oxidizing agents are incompatible (such as hydrogen peroxide and potassium permanganate). So the chemicals within each class must be further divided into chemically related and compatible groups. *Chemical Safety in the Laboratory* offers compatibility lists for some of the more commonly used laboratory chemicals. Some examples from these lists are given here.

Examples of Inorganic Compatible Groups

- acids (except nitric acid)
- amides, azides, nitrates (except ammonia gas or aqueous ammonia)
- halides, halogens
- borates, chromates, manganates, permanganates

Examples of Organic Compatible Groups

- acids, anhydrides, peracids
- alcohols, glycols
- ethers, ketenes, ketones
- nitriles, sulfides, sulfoxides

Chemical Compatibility and Storage Containers

Another consideration for chemical compatibility is potential interaction between a chemical and the material it is stored in, piped through, or otherwise in contact with. Although chemicals should always be shipped from the manufacturer in containers made of compatible materials, chemicals are often transferred into different containers for laboratory use. Selecting containers made of appropriate materials requires knowing how the chemical to be stored reacts with the materials that containers might be made of, such as glass, various plastics, or metals. For example, acetic acid has no effect on low-density polyethylene (LDPE) but will degrade polyvinyl chloride (PVC). Acetylene has no effect on aluminum but reacts with copper to form insoluble explosive compounds. A good source for information on the compatibility of chemicals and storage containers is a chemical resistance chart available free upon request from the Cole-Parmer Instrument Company (800/323-4340) or as a searchable database at their website, *www.coleparmer.com/chemcomp/icrd2.htm.*

Note that certain chemicals should not be transferred from their original containers for storage, as these original containers are designed to alleviate problems with decomposition products that may form over time or to keep the reagent away from air or airborne species with which it would react. For example, a 30% hydrogen peroxide bottle is designed for automatic release of oxygen produced as it decomposes.

Section 6F: Avoiding Hazardous Chemical Combinations
Exercise

In this Exercise, you will use the Chemical Reactivity Worksheet, a free computer program, to find out about the reactivity of substances or mixtures of substances. The Worksheet incorporates a database of reactivity information for more than 4,000 common hazardous chemicals. The database includes information about the special hazards of each chemical and about whether a chemical reacts with air, water, or other materials.

The Chemical Reactivity Worksheet ("Reactive.sta" and supporting files) is an Oracle Media Objects™ stack that runs on Macintosh computers, as well as in Microsoft Windows. To run, it needs the runtime application "OMOPlayer" and supporting files, which are distributed with the Worksheet.

To download the Worksheet, you can use a Web browser such as Netscape Navigator or Microsoft Internet Explorer. In your browser, open the location *response.restoration.noaa.gov/chemaids/react.html* and follow the instructions you see on your screen.

The Reactivity Worksheet installer program has been compressed so that you can download it from the Internet. Files are offered in a .sit format for Macintosh and an .exe format for Windows. After downloading, you need to uncompress the files. The Windows .exe files will automatically self-uncompress when double-clicked. If you are working on a Macintosh, you need a decoding utility to install the Reactivity Worksheet. (To obtain a free copy of the decoding utility Stuffit Expander, in your Web browser, open the location *www.aladdinsys.com/expander/index.html* and follow the directions you see on your screen.) To uncompact the installer program, use your decoding utility. (If you are using Stuffit Expander, either (1) drag the file onto the Stuffit Expander icon on your desktop, or (2) start Stuffit Expander, then from its File menu, choose Expand, then select the file.)

Once you have uncompressed the Reactivity Installer, double-click on it, click Install after viewing the introductory screens, and then follow the directions you see on your screen. The installer will look for a folder named "Hazmat." If it doesn't find this folder, it will create a new "Hazmat" folder on your hard drive (or startup volume, if you have more than one hard drive). The installer creates a "Hazmat" directory (folder) at the location you indicate on your hard drive and places the Reactivity Worksheet and its supporting files in that directory. To ensure trouble-free operation of the Worksheet, do not move any files out of the "Hazmat" directory or rename any files it contains.

Open the "Hazmat" folder, and then double-click the file "Reactive.sta" to start the Reactivity Worksheet. Follow the directions on your screen to learn how to use it.

The chemical pairs listed below could be very close to each other in a stockroom that is organized alphabetically, creating a potential mixing hazard. Use the Chemical Reactivity Worksheet to record the reactivity group of each chemical in the pair and list the possible reactions between the two chemicals. Be sure to enter the chemical names exactly as they appear within the quotation marks.

1. "acetic acid, [aqueous solution]" and "acetaldehyde"
   – *carboxylic acid and aldehyde*
   – *heat generation by chemical reaction; may cause violent polymerization, possibly with heat/toxic or flammable gas generation with explosive reaction; causes pressurization*

2. "aluminum powder, [metallic]" and "ammonium nitrate fertilizer"
   – *active metal and inorganic oxidizing/reducing agent*
   – *explosive when mixed with oxidizing substances; fire from exothermic reaction; ignition of products or reactants; heat generation by chemical reaction*

3. "barium" and "carbon tetrachloride"
   – *active metal and halogenated organic compound*
   – *forms very unstable metallic compounds; heat generation by chemical reaction*

4. "carbon, [activated]" and any bromate, chlorate, or iodate
   – *inorganic reducing agent and inorganic oxidizing agent*
   – *heat generated from chemical reaction may initiate explosion; may cause fire*

5. "copper" and any bromate, chlorate, or iodate
   – *less reactive metal and inorganic oxidizing agent*
   – *may cause fire; heat generation by chemical reaction*

6. "ferrous sulfate" and "hydrogen peroxide solution, [40% to 52%]"
   – *inorganic reducing agent and inorganic oxidizing agent*
   – *heat generated from chemical reaction may initiate explosion; may cause fire*

7. "lithium metal" and "maleic anhydride"
   – *alkali metal, very active acid anhydride*
   – *explosive due to vigorous reaction or reaction products; may cause fire*

8. "nitric acid, [> 40%]" and "phosphorus, [amorphous, red]"
   – *acid/inorganic oxidizing agent and inorganic reducing agent*
   – *explosive when mixed with oxidizing agent; fire from exothermic reaction; ignition of products or reactants; flammable gas generation; can become highly flammable in use; causes pressurization*

9. "potassium, [metal]" and "potassium peroxide"
   – *alkali metal, very reactive and base/inorganic oxidizing agent*
   – *explosive when mixed with combustible material; heat generated from chemical reaction may initiate explosion; may cause fire; flammable gas generator*

Section 6G: Categorizing Unknown Wastes for Disposal
Instructor Notes

The objective of this section is for students to learn how to characterize unknown laboratory wastes for disposal. In a paper-and-pencil exercise, students use a decision-tree flowchart from *Prudent Practices in the Laboratory* to determine the appropriate hazard category for unknown waste chemicals so that appropriate disposal procedures can be followed.

Student Pages

*   Section 6G: Categorizing Unknown Wastes for Disposal—Background
*   Section 6G: Categorizing Unknown Wastes for Disposal—Exercise

Possible Playout of the Section

Students read the Background and complete the Exercise.

Section 6G: Categorizing Unknown Wastes for Disposal

Background

Note: This section provides general recommendations for and information about categorizing unknown wastes for disposal. However, you should always follow the procedures established by your organization's chemical hygiene plan.

In general terms, waste is defined as surplus, unnecessary, or unwanted material. All proper waste disposal takes one of three forms: into the atmosphere, through evaporation or incinerator emissions; into waterways via sewer systems and wastewater treatment facilities; or into landfills. No matter what method of disposal is used, waste in some way affects people other than the ones who acquired or produced it. Therefore, those who generate waste in the laboratory have a responsibility to consider what will happen to that waste after it leaves the laboratory. Important factors to consider include disposal cost (for many materials, this cost can be high), potential hazards posed to people, and the environmental impact of the waste.

As a laboratory worker, you will be the first person responsible for proper waste disposal, since you will be in the best position to know the identities and characteristics of the materials used and produced. In choosing a method of handling and minimizing the waste, you will need to consider the hazards and risks of the waste produced. Some general recommendations follow:

- *All waste containers must be labeled:* Any unknown waste must be identified, a costly and time-consuming procedure.

- *Do not mix hazardous and nonhazardous waste:* Once mixed, all the waste is then considered hazardous and subject to expensive disposal procedures.

- *Keep chemically different hazardous wastes separate:* If wastes must be disposed of off-site, more options are available if the various types of waste are kept separate.

How Is Waste Disposal and Cleanup Regulated?

In the United States, disposal of chemically hazardous waste is governed by the Resource Conservation and Recovery Act (RCRA) of 1976. Cleanup of wastes disposed of unsafely in the past is governed by the Comprehensive Environmental Response, Compensation, and Liability Act (CERCLA) of 1980, more commonly known as "Superfund." CERCLA also lists wastes that cannot be disposed of in landfills. RCRA establishes legal definitions for hazardous wastes and sets guidelines for managing, storing, transporting, and disposing of them, while CERCLA applies to wastes disposed of unsafely in the past. RCRA and CERCLA are federal regulations administered by the federal Environmental Protection Agency (EPA); some states and local jurisdictions have additional, often stricter, regulations.

According to federal regulations, hazardous wastes include chemical compounds that are on one of several regulatory lists (listed wastes) or that are ignitable, corrosive, reactive, or toxic. Federal lists and regulations are updated from time to time, so waste generators should always obtain detailed, up-to-date information. In general, all waste chemicals should be considered hazardous unless good reasons exist for considering them nonhazardous. Radioactive waste and infectious waste are two other categories of waste regulated by law. Disposal, transport, and storage of radioactive waste is regulated by the U.S. Nuclear Regulatory Commission (U.S. NRC). Infectious waste is generally regulated by state law.

Familiarity with both RCRA and CERCLA regulations is essential for people responsible for disposing of waste. In some cases, chemicals listed under one act (RCRA or CERCLA) are not listed under the other, and vice versa. Since institutions are still liable for the waste they generate even after it has been disposed of, compliance with both sets of regulations is critical. The waste generator is always liable for any violations in waste disposal, even if the waste has been disposed of by a contractor.

How Can Hazardous Waste Be Disposed Of?

Hazardous waste may not be simply poured down the drain or thrown in an ordinary garbage can. Instead, the waste must be disposed of in a way that reduces or eliminates its impact on human health and the environment. Landfilling in a secure landfill (a landfill designed to handle hazardous waste) is one option, but it is very expensive, and secure landfills are being closed faster than they are being replaced. Incineration is becoming the most common disposal method, but environmental concerns and the difficulty of obtaining permits for new commercial incineration facilities put the long-term future of this option in doubt.

Because of these disadvantages of traditional disposal methods, waste minimization has become an important strategy. Waste minimization includes using smaller quantities of chemicals, arranging chemical exchanges between laboratories, keeping hazardous and nonhazardous waste products separate from each other to prevent contamination of the nonhazardous waste, and clearly defining the line of responsibility for chemical waste management. Depending on the resources available in the laboratory, some hazardous wastes may also be treated in-house to reduce or eliminate their hazards. However, except for elementary neutralization, procedures for treating hazardous waste may require a permit unless the treatment technique is part of the experimental protocol. Under RCRA, a large-quantity generator (a facility producing more than 1,000 kg of hazardous waste per month) may not store hazardous waste for more than 90 days unless it is a permitted treatment, storage, or disposal facility of hazardous waste. Small-quantity generators (facilities producing between 100 and 1,000 kg per month) may not store hazardous waste for more than 180 days.

The specific way in which a particular waste material is handled and disposed of depends on what hazard it poses. However, waste doesn't always fit neatly into one category of hazard. Multihazardous wastes (referred to as mixed waste) pose special problems, because the treatment method for one hazard may be inappropriate for the treatment of one or more of the others. Institutions that generate multihazardous waste must work with the appropriate regulatory agencies to resolve inconsistencies and develop safe, consistent plans for management and disposal.

Disposal of Unknown Materials

Choosing a method of disposal requires information about the properties of the waste. Thus, all chemicals used or generated in the laboratory should be identified clearly both on the container and in a readily accessible record in the laboratory. Unidentified materials pose a big problem: treatment facilities are prohibited from accepting materials whose hazards are unknown.

Before waste disposal facilities will consider handling unidentified materials, they usually require the following information about the material:

- physical description
- water reactivity
- water solubility
- pH and sometimes neutralization information
- ignitability
- presence of oxidizer
- presence of sulfides or cyanides
- presence of halogens
- presence of radioactive materials
- presence of biohazardous materials
- presence of toxic constituents

While tests to determine these characteristics are simple for a trained laboratory worker to perform, they should be carried out with great caution on unknown substances. The first concern in identification should be safety. The use of PPE and safety devices, such as shields and fume hoods, is absolutely essential.

Tests on unknown waste chemicals should be carried out according to a specific flowchart, or decision tree, shown in Figure 6G-1 in the Exercise.

Section 6G: Categorizing Unknown Wastes for Disposal
Exercise

Unidentified waste materials present a problem, because treatment and disposal facilities are not allowed to accept materials whose hazards are not known. In this Exercise you will learn how simple test results can be used to determine the hazard class of unidentified waste materials. The flowchart below represents a sequence of testing that allows the laboratory to categorize the hazards of unidentified waste. "DWW" means "dangerous when wet," and "NOS" means "not otherwise specified." For the tests, (+) means the test is positive and (−) means the test is negative.

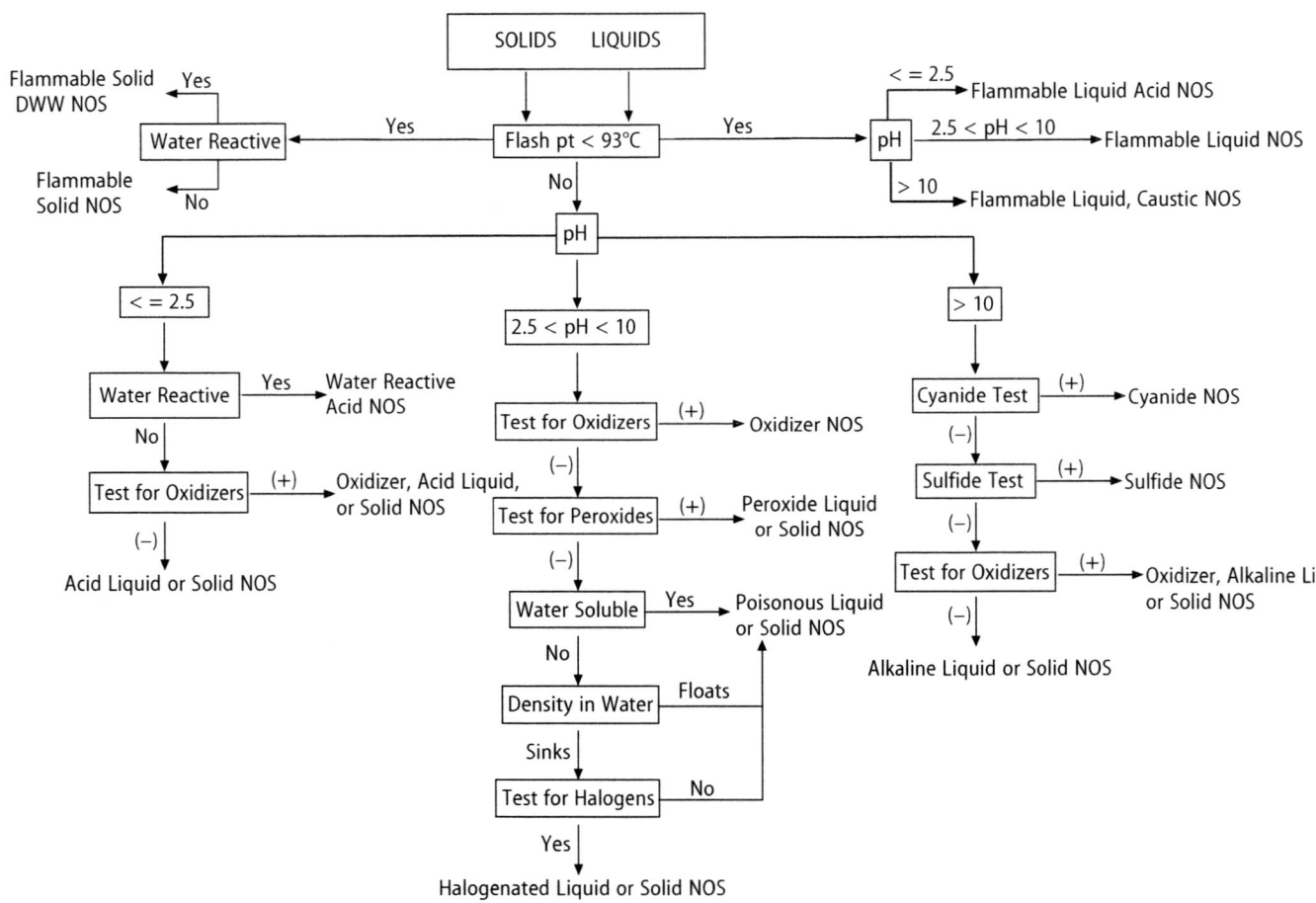

Figure 6G-1: Flow Chart for Categorizing Unidentified Chemical Waste
(Source: Prudent Practices, reproduced with permission)

Using the flowchart and descriptions of the test outcomes for four unidentified waste chemicals listed below, determine the waste category for each.

1. This material is a solid with a flash point of 104°C and a pH of 2. It does not react with water and tested positive in the oxidizer test.
    oxidizer, acid solid NOS

2. This material is a solid with a flash point of 81°C, and it does not react with water.
    flammable solid NOS

3. This material is a liquid with a flash point of 53°C and a pH of 7.
    flammable liquid NOS

4. This material is a liquid with a flash point of 120°C and a pH of 11. It tested negative in the cyanide, sulfide, and oxidizer tests.
    alkaline liquid NOS

Chapter 7
Emergency Equipment

During an experiment on measuring temperature, alcohol lamps were being used by students. One of the lamps must have had a hairline crack (not visible). When the lamp was lit, it exploded, spewing alcohol all over the table. The students were sent from the room. When I went for the fire extinguisher it was empty. I cleared the desk and surrounding desks of books and materials. I went for the fire blanket—it was just a regular blanket. When the alcohol burned itself out, we returned to the room. Fortunately no one was hurt and no damage occurred except to the "not real" fire blanket.

Learning by Accident, page 10

Chapter 7 Objectives

Pertinent Voluntary Industry Standards

- *VIS L2 Standards*—Choose the proper safety equipment for conducting a variety of laboratory tasks (e.g., proper hoods, shields).

- *VIS (L2.04):02*—Demonstrate the appropriate use of safety equipment, including but not limited to safety glasses, showers, respirators, eye washes, and blankets.

- *VIS (L2.01):12*—Read a variety of cleanup and emergency response procedures and determine how to implement the procedures.

- *VIS (L2.01):13*—Describe procedures used to respond to a spill or release.

Goals of the Chapter

Students will learn

- how to respond to laboratory emergencies and correctly use equipment such as fire extinguishers, fire blankets, eyewashes, and deluge showers; and

- how to inspect emergency equipment for safe operation.

Why Your Students Need to Understand Emergency Equipment

Most people have never experienced a life-threatening situation, such as an explosion or a chemical spill. The closest many have come to a real emergency is seeing a video or watching a disaster film on television or in a theater. However, odds are that your students will be faced with several emergencies in their careers.

When a true emergency does occur, it often catches us by surprise. Our first feelings are usually ones of fear. If we "lose our heads" and begin to panic, we may respond improperly during an emergency. Knowledge, training, and frequent drills reinforce how we should properly respond to an emergency. All this preparation helps us to suppress our own panic and allows us and others to react appropriately. Plus, we gain control over the situation until professional firefighters, paramedics, and other officials arrive on the scene to take over. However, learning how to act in emergency situations takes practice.

Overview of the Chapter

Chapter 7: Emergency Equipment contains four sections. General background information and complete instructions for the exercises are provided. The background pages are intended only to provide basic background information on emergency procedures for your students. We recommend that you extend this background information by having students find and read the pertinent sections of the books in the safety library (described in Chapter 1). You may wish to have students briefly report on the information they find during the monthly safety meetings.

Section 7A: Getting to Know Emergency Equipment

This section introduces students to the importance of emergency equipment in the laboratory and workplace. To initiate interest in the topic, students read and discuss Laboratory Incident Reports concerning accidents that could have been ameliorated through the availability and proper use of emergency equipment.

Section 7B: Responding to Fires

The objective of this section is for students to understand the types of fires that can occur in a laboratory and the types of fire extinguishers needed to extinguish them. In the Exercise, students learn basic fire response strategies and practice using a fire extinguisher.

Section 7C: Eyewash Stations and Deluge Showers

In this section, students practice finding eyewash stations with their eyes closed and time each other getting to the deluge showers. Students also practice routine checks of eyewash stations and showers. The objective of this section is for students to understand when and how to use eyewash stations and deluge showers and how to check them for safe operation. Students learn that in a laboratory situation, a splash on the eye or skin requires immediate attention, with no time for panic or confusion about treatment. Proper treatment also requires fully functioning emergency equipment.

Section 7D: Responding to Spills

The objective of the section is to prepare students to use spill kits appropriately in response to spills of various hazardous materials. Students practice spill-response techniques using vinegar as a substitute for more corrosive materials.

References for Chapter 7

The ACS Guide to Spill Response Planning in Laboratories; American Chemical Society: Washington, DC, 1997.

EM Science Webpage. MSDS for Nitric Acid, 65%. www.emscience.com/search/index.asp (accessed Jan 25, 2000).

Emergency Eyewash and Shower Equipment. American National Standards Institute, ANSI Z358.1-1998.

"Fire Extinguishers: Safety In the School Laboratory," *Flinn Fax!.* 1986, *Vol. 86-1.*

Flinn Chemical & Biological Catalog Reference Manual, Flinn Scientific: Batavia, IL, 2000; p 894.

Gorman, C. *Working Safely with Chemicals in the Laboratory,* 2nd ed.; Genium: Schenectady, NY, 1997.

"A Guide to Portable Fire Extinguishers"; Ohio Department of Commerce, Division of State Fire Marshal: Reynoldsburg, OH, 1995.

Hazard Communication Standard. *Code of Federal Regulations,* 29 CFR 1910.1200, 1999. (This is available free online at www.osha.gov.)

Hofstader, R.; Chapman, K. *Foundations for Excellence in the Chemical Process Industries: Voluntary Industry Standards for Chemical Process Industries Technical Workers;* American Chemical Society: Washington, DC, 1997.

Improving Safety in the Chemical Laboratory: A Practical Guide, 2nd ed.; Young, J.A., Ed.; John Wiley & Sons: New York, 1991; pp 94–100.

Learning By Accident; Mojtabai, F.; Kaufman, J., Eds.; Laboratory Safety Institute: Natick, MA, 1997.

Learning By Accident, Vol. 2; Mojtabai, F.; Kaufman, J., Eds.; Laboratory Safety Institute: Natick, MA, 2000.

Medical Services and First Aid. *Code of Federal Regulations,* 29 CFR 1910.151, 1999. (This is available free online at www.osha.gov.)

National Research Council Recommendations Concerning Chemical Hygiene in Laboratories (Non-Mandatory). *Code of Federal Regulations.* 29 CFR 1450.1910, Appendix A, 1998. (This is available free online at www.osha.gov.)

Occupational Exposure to Hazardous Chemicals in Laboratories Standard. *Code of Federal Regulations,* 29 CFR 1910.1450, 1999. (This is available free online at www.osha.gov.)

Portable Fire Extinguishers. *Code of Federal Regulations,* 29 CFR 1910.157, 1999. (This is available free online at www.osha.gov.)

Prudent Practices in the Laboratory; National Academy: Washington, DC, 1995; pp 85–86, 89, 134.

Safety in Academic Chemistry Laboratories, 6th ed.; American Chemical Society: Washington, DC, 1998.

Segal, E. "First Aid for a Unique Acid, HF: A Sequel," *Chemical Health & Safety.* 2000, *7* (1), 18–23.

US Department of Energy. *Potentially Hazardous Amoebae Found in Eyewash Stations;* EH-15; Department of Energy: Washington, DC, 1986.

Acknowledgments

Images of deluge shower, eyewash, and eyewash gauge provided by Haws Corporation, Berkeley, CA.

Section 7A: Getting to Know Emergency Equipment
Instructor Notes

This section introduces students to the importance of emergency equipment in the laboratory and workplace. To initiate interest in the topic, students read and discuss Laboratory Incident Reports concerning accidents that could have been ameliorated through the availability and proper use of emergency equipment.

Student Pages

- Section 7A: Getting to Know Emergency Equipment—Laboratory Incident Reports
- Section 7A: Getting to Know Emergency Equipment—Background

Possible Playout of the Section

Students read the Laboratory Incident Reports and Background and answer the following questions.

1. What are similarities and differences between the incidents described in the Laboratory Incident Reports?

2. What conclusions can you draw about safety from these incidents?

3. How could each of these accidents have been prevented?

Section 7A: Getting to Know Emergency Equipment
Laboratory Incident Reports

Incident Description: Eyewashes and Deluge Showers

At a midwest university, an untrained technician was carrying one gallon of chromic acid cleaning solution. The container struck the corner of a lab bench and shattered. No safety shower was present. She was hospitalized for three weeks and required skin grafts.

Learning by Accident, page 15

A senior taking a chemistry course had his last lab of the week on Friday from 2–5 p.m. The lab called for heating benzene in a water bath. The student was in a hurry and heated with a Bunsen burner. Vapors overflowed, ignited, trailed back and blew up in his face. The student ran to the shower—it did not work. The fire blanket fell apart. His goggles melted. Finally the fire went out.

Learning by Accident, page 12

Incident Description: Fire Extinguishers

During an experiment on measuring temperature, alcohol lamps were being used by students. One of the lamps must have had a hairline crack (not visible). When the lamp was lit, it exploded, spewing alcohol all over the table. The students were sent from the room. When I went for the fire extinguisher it was empty. I cleared the desk and surrounding desks of books and materials. I went for the fire blanket—it was just a regular blanket. When the alcohol burned itself out, we returned to the room. Fortunately no one was hurt and no damage occurred except to the "not real" fire blanket.

Learning by Accident, page 10

Incident Description: Spill

A substitute teacher with no science preparation had the class building kites. The kites were stored in the storeroom. The substitute permitted students to retrieve the kites without supervision. A student stepped on the bottom shelf to reach the kites. The shelf support dislodged, the shelf dropped, and bottles of ammonium hydroxide, glacial acetic acid, and isopropyl alcohol fell to the floor. Dense fumes formed. The fire department was called. Hose lines were run to the room on the second floor, and smoke evacuation fans were used.

Learning by Accident, Vol. 2, page 11

Section 7A: Getting to Know Emergency Equipment
Background

Note: This handout provides general recommendations for and information about emergency equipment. However, you should always follow the emergency plan established by your organization's chemical hygiene plan.

Even when you follow good laboratory safety practices, accidents and emergencies can still happen. Part of good safety practice is being prepared for emergencies. When an accident happens to you or a co-worker in the lab, you may have just seconds to begin using emergency equipment to minimize injury and property damage. Acting inappropriately because you do not know emergency procedures can make the consequences of an emergency worse. The chemical hygiene plan required by your organization under OSHA's Laboratory Standard should have detailed descriptions of all emergency procedures. *Prudent Practices in the Laboratory* states that you should not undertake laboratory work without knowledge of the following points:

- how to report a fire, injury, chemical spill, or other emergency to summon emergency response;

- the location of emergency equipment such as deluge showers, first-aid kits, and eyewashes;

- the location of fire extinguishers, fire blankets, and other fire equipment; and

- the locations of all available exits for evacuation from the laboratory.

Of course all emergency equipment is useless if you are unconscious or incapacitated. The most important safety asset you can have during an emergency is the help of co-workers or laboratory partners. **Never work alone in a laboratory.**

Types of Emergency Equipment in the Laboratory

Emergency equipment in the laboratory is designed to aid your response to fires, chemical splashes in eyes or on skin, and accidental chemical releases or spills. Three general types of emergency equipment are found in laboratories:

- Fire extinguishers can help you put out a small fire before it spreads. But safely using a fire extinguisher requires knowledge of the right extinguisher for the type of fire, and prior hands-on experience putting out fires is beneficial. *You will learn fire response strategies, including selecting and using fire extinguishers, in Section 7B: Responding to Fires.*

- Deluge showers and eyewashes provide a sustained source of clean water for flushing contaminated eyes and skin. *You will learn about inspecting, testing, and using deluge showers and eyewashes in Section 7C: Eyewash Stations and Deluge Showers.*

- Spill kits include materials to contain, neutralize, and absorb spills. Different spill kits are available for common types of hazardous spills, such as liquids, acids, bases, mercury, ammonia, and blood and other potentially infectious materials. *You will learn how to clean up a hazardous spill in Section 7D: Responding to Spills.*

- While first-aid kits provide bandages and antibacterial solutions, nothing should be applied to a burn except a sterile gauze dressing. Also, note that while laboratory first-aid kits may also contain an eyewash bottle, such a bottle is NOT an acceptable substitute for an eyewash fountain.

Regulations Governing Emergency Equipment

When a co-worker's clothing is splashed with a concentrated acid, you don't want to discover that the deluge shower doesn't work. OSHA regulations require laboratories to develop a chemical hygiene plan to prevent such situations from happening. Under the Laboratory Standard, an outline of recommended concerns briefly discusses the care, use, and availability of emergency equipment. Many laboratories use ANSI standards as a guideline for their chemical hygiene plan because these standards give more detail and are often more stringent than OSHA's recommendations. ANSI standards explicitly state the number of safety signs required, how accessible emergency equipment should be in a laboratory, how often the equipment should be tested, and what the equipment's operating requirements are.

General Response to Emergencies

In any emergency, follow your organization's established procedures. Be aware of, practice, and follow established procedures for evacuation. The suggestions given below, adapted from *Prudent Practices in the Laboratory*, are intended to provide general guidelines:

- Notify the laboratory personnel and supervisor of the accident or fire and, if necessary, evacuate the area.

- Tend to any injured or contaminated personnel and, if necessary, request help.

- Take steps to confine and limit the spill or fire if these steps can be done without risk of injury or contamination.

Section 7B: Responding to Fires
Instructor Notes

The objective of this section is for students to understand the types of fires that can occur in a laboratory and the types of fire extinguishers needed to extinguish them. In the Exercise, students learn basic fire response strategies and practice using a fire extinguisher.

Student Pages

- Section 7B: Responding to Fires—Background
- Section 7B: Responding to Fires—Exercise

Possible Playout of the Section

Students read the Background and do the Exercise on matching proper fire extinguishers to fires. Emphasize to students that firefighting information they will be researching in the Exercise is for instructional purposes only and is not intended as advice for putting out actual fires. In case of fire, they should follow fire procedures established by their organization's chemical hygiene plan, as well as local and state ordinances.

Because fire professionals strongly recommend that persons working with hazardous materials learn how to use fire extinguishers properly before a fire occurs, you may wish to follow up the student exercise with a fire extinguisher class. This class can typically be set up through your university's safety officer, a local fire marshal, or a fire extinguisher manufacturing company. (Some agencies may charge a nominal fee for the class.) OSHA regulations require employers to provide an educational program to familiarize employees with the general principles of fire extinguisher use and the hazards involved with incipient-stage firefighting (Fire and Emergency Plans 1910.157).

A fire extinguisher class is most effective if the students can practice on a live fire. Setting such a fire requires an open burn permit. Procedures for obtaining a permit vary; contact your local fire marshal or university safety officer for guidance. The extinguisher class instructor will also inform you whether you need to provide a burn tray (a low-profile container made of organic material such as paper, dried wood, or excelsior) and fire extinguishers for the class. Usually, the extinguisher class instructor will provide all necessary materials.

Safety

The labs in this book do not routinely include special safety precautions for the chemicals or equipment that might be used. As the course instructor, it is expected that you will provide students with access to SOPs, MSDSs, and other resources they need to safely work in the laboratory while meeting all regulatory requirements. Before doing any of the labs from this book or from other sources, we recommend that you regularly review special handling issues with students, allow time for questions, and then assess student understanding of these issues.

Setup Notes

Extinguishers can often be provided by your safety officer, fire marshal, or campus maintenance personnel. Also, extinguishers nearing their recharge date may be available from local industry. Pressurized-water fire extinguishers are NOT recommended especially in a laboratory environment. ABC fire extinguishers are preferable because of their versatility.

Section 7B: Responding to Fires
Background

Note: This handout provides general recommendations for and information about emergency equipment. However, you should always follow the emergency plan established by your organization's chemical hygiene plan.

Fires are some of the most common laboratory accidents and can occur despite your best efforts to handle chemicals and laboratory equipment safely. Responding to fires requires quick action and an awareness of many factors. Understanding the conditions necessary for a fire to start will help you understand how you can help put it out. As discussed in the Background for Chapter 6, Section 6C, fires must have an ignition source, fuel, and an oxidizer in order to start; removing the fuel or the oxidizer will put out the fire.

Firefighting Measures and MSDSs

If a chemical fire starts in a laboratory where you work, familiarity with the information in Sections 3–6 of the MSDS for the chemical(s) involved will be imperative for everyone working in the laboratory and everyone responding to the fire. (See excerpts from the nitric acid MSDS below and on the next page.) Although Section 5: Firefighting Measures directly addresses fighting the fire, persons who are exposed to fumes may require first aid (Section 4), and accidental releases (Section 6) may occur due to destruction of containers during the fire. Note that Section 5 describes the recommended extinguishing media, any hazardous products of combustion, and possibly hazardous reactions. It is imperative to read and understand Section 5 before working with a chemical. There is no time to read the MSDS sheet after the fire has started.

MSDS Excerpts for Nitric Acid (65% HNO_3) Sections 3–6

<u>Section 3: Hazards Identification</u>

Emergency Overview
CAUSES SEVERE BURNS.
VAPOR EXTREMELY HAZARDOUS.
May cause nitrous gas poisoning.
MAY BE FATAL IF INHALED OR SWALLOWED.
Symptoms of lung injury may be delayed.
STRONG OXIDIZER
Appearance: Colorless liquid; acrid odor
Potential Health Effects (Acute and Chronic)
Symptoms of Exposure: Causes severe burns on contact with any body tissue. Inhalation of vapors or mists can cause severe burns to respiratory passages, pneumonia and pulmonary edema. Can be fatal if inhaled or swallowed. Symptoms of lung injury may be delayed.
Medical Cond. Aggravated by Exposure: Respiratory conditions
Routes of Entry: Inhalation, ingestion or skin contact.
Carcinogenicity: The material is not listed (IARC, NTP, OSHA) as cancer causing agent.

Section 4: First Aid Measures

Emergency First Aid: GET MEDICAL ASSISTANCE FOR ALL CASES OF OVEREXPOSURE.
Skin: Immediately flush thoroughly with large amounts of water.
Eyes: Immediately flush thoroughly with water for at least 15 minutes.
Inhalation: Remove to fresh air; give artificial respiration if breathing has stopped.
Ingestion: Do not induce vomiting; if conscious, give water freely and get medical attention.

Section 5: Firefighting Measures

Flash Point (F): Noncombustible
Flammable Limits LEL (%): N/A
Flammable Limits UEL (%): N/A
Extinguishing Media: Water spray, dry chemical.
Firefighting Procedures: Wear self-contained breathing apparatus and protective clothing.
Fire & Explosion Hazards: Can react explosively with certain reducing agents and combustibles; such as metal powders, carbides, H_2S, turpentine.

Section 6: Accidental Release Measures

Spill Response: Evacuate the area of all unnecessary personnel. Wear suitable protective equipment listed under Exposure/Personal Protection. Eliminate any ignition sources until the area is determined to be free from explosion or fire hazards. Contain the release and eliminate its source, if this can be done without risk. Take up and containerize for proper disposal as described under Disposal. Comply with Federal, State, and local regulations on reporting releases. Refer to Regulatory Information for reportable quantity and other regulatory data.

EM SCIENCE recommends SPILL-X neutralizers and absorbent agents for various types of spills. Additional information on the SPILL-X products can be provided through the EM SCIENCE Technical Service Department (609) 423-6300. The following EM SCIENCE SPILL-X neutralizer and absorbent is recommended for this product: SX0861 Spill-X-A Acid Spill Treatment Kit.

First Response to Fires

Most firefighting professionals recommend that your first reaction to a fire should be to call for emergency help by sounding an alarm and/or calling the fire department. They suggest that you do not begin fighting the fire before calling for help, because even if the fire seems small and manageable, it can quickly get out of control.

In case of fire, send someone to call 911. If you successfully put out the fire, simply call again and report that the fire is out. The fire department will send a small unit to make sure the fire is completely extinguished. Depending on the regulations in your state, you may be required to report any fire, even a small fire that was extinguished without the help of the fire department.

You should always be aware of the correct number to dial for an emergency. Know ahead of time whether you can dial 911 directly or whether you need to dial another number first to get an outside line. You should also know the location and use of all emergency fire equipment in your laboratory, including fire extinguishers, fire blankets, and appropriate extinguishing materials for small metal fires. If you are able, designate someone to call for help and decide whether anyone present is qualified to fight the fire. Tell everyone else to leave the area. This requires prior training in emergency evacuation procedures.

Suffocating Fires Without a Fire Extinguisher

Fires that start in a small container can usually be suffocated by covering the vessel loosely with another container such as a watch glass or beaker. Do not use dry towels or cloths that would be likely to ignite. Never try to move a vessel that contains a fire.

Fire blankets can also be used to suffocate a small fire, but do not use them on a burning person; they tend to hold in heat and increase the severity of burns. The old recommendation for using a fire blanket was to wrap the blanket around the person. This must not be done because it creates a chimney effect, causing severe upper-body and head damage. If your clothing or a co-worker's clothing catches fire, it is important not to panic. Immediately stop, drop to the floor, and roll until the flames have been put out. Alternatively, you may be able to douse the fire using the safety shower if you are very close to it.

Appropriate dry materials can be used to extinguish small metal fires. Depending on the metal, various materials can be appropriate, including sand and sodium chloride. However, fire extinguishers made explicitly for suffocating metal fires are available and are discussed in "Use the Correct Fire Extinguisher."

When to Use a Fire Extinguisher

Sometimes your safety or that of your fellow workers may depend on using an extinguisher to buy the few minutes needed for safe evacuation. Therefore, it is important to be prepared. You should consider several factors as you decide whether you are able to fight a fire in the laboratory. For example, have you had hands-on experience using different kinds of fire extinguishers? Do you know what started the fire and whether the extinguisher that is available to you is the proper size and type to use? (See the next section, "Use the Correct Fire Extinguisher.") Is the extinguisher in your laboratory in good working order? (Fire extinguishers require routine inspection and maintenance. Most laboratory and workplace fire extinguishers will have a tag showing a record of the inspection dates. Ask your fire department about local regulations.) Is there an exit from the room that is easily accessible, even if the area of the fire increases? Is the fire extinguisher located near the exits, thus preventing people from being trapped by the fire when retrieving the extinguisher? Is the fire small enough that it can be controlled with the fire extinguisher? Many people are not aware of how quickly the contents of a fire extinguisher are expended. A 10-pound CO_2 or dry chemical extinguisher may last only about 10 seconds!

Some very small confined fires are better left to burn themselves out. Some fires can be covered with a large beaker to cut off the supply of oxygen or can be smothered with a wet towel to lower the temperature of the blaze below the kindling temperature. In some cases, using a fire extinguisher is actually more dangerous than the above options. The force of the extinguishing agent can push the flame, glassware, or burning material several feet from the "safe site" into other flammable materials.

If no one in the laboratory is appropriately confident in his or her ability to use fire extinguishers, then everyone should evacuate the building and let professional firefighters handle it. If possible, have someone shut all windows and doors as everyone leaves the room in order to contain the fire, unless it would be dangerous to try to reach the windows.

Use the Correct Fire Extinguisher

Fire extinguishers can be a useful defense when fighting small fires or buying time for safe evacuation as long as you use the correct extinguisher for the type of fire you are facing. The National Fire Protection Association (NFPA) has classified four types of fires, each with its own symbol that is displayed on the fire extinguisher suitable for that type of fire (as shown in the following table).

Classes of Fires and Their Symbols			
Class	Symbol	Symbol Color	Type(s) of Fire
A	A	Green	routine combustibles such as wood, paper, cloth, rubber, and plastics
B	B	Red	flammable liquids and flammable gases
C	C	Blue	electrical equipment
D	D	Yellow	flammable metals

Using the wrong type of extinguisher may pose more of a threat than using nothing at all and waiting for the fire department to come. Multipurpose extinguishers are the most popular because they eliminate much of the decision-making in selecting the correct extinguisher. However, multipurpose extinguishers should never be used on a Class D fire involving flammable metals. You should know the types of fire extinguishers available and the types of fires that they can be used to extinguish. Before using any fire extinguisher, always check the label to make sure it is appropriate for the class of fire you have.

Water extinguishers are effective on Class A fires only. They should never be used on Class B, C, or D fires. Water extinguishes Class A fires because its high heat capacity cools the fuel and it excludes oxygen by forming steam, thereby removing two legs of the fire tetrahedron. Since water extinguishers can spray over long distances, the firefighter is farther from the flames and therefore safer. Using water on a Class B fire, however, can make the fire worse, as most flammable organic liquids are less dense than water. The burning fuel will merely float on top of the water and continue to spread. Since water is an electrical conductor, it should never be used on Class C fires. Water should also never be used on Class D fires because of the potential chemical reactions that may occur between the water and the burning metal.

Carbon dioxide (CO_2) extinguishers are effective on Class B and C fires because the gas both smothers and cools the fuel. Because they leave no residue, CO_2 extinguishers are especially recommended for fires involving computers or delicate equipment. However, the range of a CO_2 extinguisher is only about 3–8 feet, which means the firefighter must stand close to the flames. These extinguishers should not be used in closed spaces because of their short range and the fact that CO_2 gas is a simple asphyxiant. The contents of a CO_2 extinguisher are under great pressure and are expelled with such force that they can scatter Class A fuels, like wood and paper, or tip over burning containers without extinguishing them. Like water extinguishers, CO_2 extinguishers should never be used on Class D fires because of the potential chemical reactions that may occur. The gas is delivered at extremely cold temperatures and may cause thermal shock and/or cracking in electrical circuits and can cause frostbite if sprayed on the human body. Care must also be taken when holding the nozzle of a CO_2 extinguisher because it may get very cold.

Monoammonium phosphate (MAP) extinguishers are dry chemical powder extinguishers that can be used on Class A, B, and C fires, a diversity that has made them the most common extinguisher in homes, automobiles, and laboratories. The expelled monoammonium phosphate melts to form a sticky residue, which smothers the fuel surface and excludes oxygen. There is no cooling effect, however, and flare-ups may occur. The usual practice is to spray the area with water after the fire has been completely extinguished to cool down the area. Class C fires extinguished with MAP extinguishers should not be sprayed with water until you are certain the equipment has been de-energized. Although MAP extinguishers are highly effective, they pose a cleanup problem, especially for electronic equipment. Like water and CO_2 extinguishers, MAP extinguishers should not be used on Class D fires.

Potassium carbonate or sodium carbonate extinguishers are a second type of dry chemical powder extinguisher. Either of these two chemicals can be used on Class B or C fires but not on Class A or D fires. Potassium carbonate (K_2CO_3) and sodium carbonate (Na_2CO_3) are effective for two reasons. First, the chemicals cause the saponification—hydrolysis into soap—of oils and fats at the fuel's surface, hampering vaporization and decreasing the chance of re-ignition. Secondly, the chemicals act as free radical scavengers and interfere with the burning process, eliminating a leg of the fire tetrahedron. Cleaning up the potassium carbonate or sodium carbonate powder is troublesome and poses the same problems for electrical equipment that the MAP extinguisher does.

Class D extinguishers are specialized because of the highly reactive nature of the metals that cause Class D fires. Sand, powdered sodium chloride, graphite, and soda ash are the most common Class D extinguishers for alkali metal fires. These substances must be completely dry, however, as moisture may react with the metal. Met-L-X® is a special fire-extinguishing material that is effective on all alkali metal fires except lithium. Dry graphite is suitable for extinguishing burning lithium, but it must be completely dry; wet graphite will react with lithium to form lithium carbide, which reacts with water to form acetylene gas, a fuel.

Regardless of the type of fire you have, fire extinguishers are effective only at certain ranges and for certain periods of time. The following table shows the range and duration of discharge estimates for different volumes of the various types of fire extinguishers. Class D fire extinguishers vary in capacity, range, and duration so much that they are not included.

Range and Duration Estimates for Capacities of Extinguishers			
Type	Capacity	Range (horizontal feet)	Duration of Discharge (seconds)
Water	2.5 gallons	30–40	45
Carbon dioxide	2.5–5 pounds	3–8	8–30
	10–15 pounds	3–8	8–30
	20 pounds	3–8	10–30
Dry Chemical (monoammonium phosphate or potassium or sodium carbonate)	1–5 pounds	5–12	8–10
	9–17 pounds	5–20	10–25
	17–30 pounds	5–20	10–25

How to Use a Fire Extinguisher

If you determine that you are able to fight a laboratory fire and you have the proper extinguisher, follow NFPA's PASS procedure to put out the flames:

- **P**ull the safety pin from the top of the extinguisher.

- **A**im the nozzle or hose at the base of the flames. Realize that you may have to get quite close to the fire (as close as 3 feet with some CO_2 extinguishers).

- **S**queeze the handle.

- **S**weep from the outside edge to the center so that you do not split the fire in two, and sweep from front to back so you do not move the fire toward you.

Even the correct extinguisher for a specific type of fire can be used incorrectly. If possible remain upwind. This will increase your safety and assist in extinguishing the fire. Be sure that you keep yourself between the fire and the exit so you don't become trapped inside a burning room. Remember, the fire extinguisher must be located near the exits, thus preventing people from being trapped by the fire when retrieving the extinguisher. If your efforts to put out a fire don't work, you need to be able to escape quickly.

Always Have a Fire Emergency Plan

Fires in laboratories can ignite in a split second, so all laboratory workers should know ahead of time what they need to do if a fire occurs. In spite of the fact that the manager or coordinator of a laboratory is often responsible for ensuring that all laboratory workers know what to do in case of a fire, all workers should proactively learn this information. Necessary information includes where fire extinguishers are located, what type(s) of extinguishers are available, exit routes, and emergency phone numbers.

Section 7B: Responding to Fires
Exercise

The objective of this Exercise is for you to understand the conditions for fire, types of fire and fire extinguishers, and basic fire response strategies. MSDSs provide helpful information in this regard. The firefighting information you will be researching in this Exercise is for instructional purposes only and is not intended as advice for putting out actual fires. In case of fire, you should follow firefighting procedures established by your organization's chemical hygiene plan, as well as local and state ordinances.

1. Assume that the following materials are either on fire or are in close proximity to a fire. Indicate for each the appropriate extinguishing media (if the material is on fire), and any fire/explosion hazards that may exist.

    The answers provided below come from the Firefighting sections of MSDSs provided by Fisher Scientific (www.fishersci.com), Chevron (www.library.cbest.chevron.com), and MG Industries (www.mgindustries.com).

    a. Nitric acid, 90%, CAS #7697-37-2
        From Fisher Scientific

        General Information: As in any fire, wear a self-contained breathing apparatus in pressure demand, MSHA/NIOSH (approved or equivalent), and full protective gear. Strong oxidizer. Contact with combustible materials may cause a fire. Use water spray to keep fire-exposed containers cool. Substance is noncombustible. Vapors may accumulate in confined spaces. Contact with metals may evolve flammable hydrogen gas. Substance may react with water, and may release corrosive and/or toxic gas. Extinguishing Media: Substance is noncombustible; use agent most appropriate to extinguish surrounding fire. Do NOT use straight streams of water. Cool containers with flooding quantities of water until well after fire is out.

    b. Methane, CAS #74-82-8
        From MG Industries

        FIRE AND EXPLOSION HAZARDS: Severe fire hazard. Severe explosion hazard. Containers may rupture or explode if exposed to heat. Vapor/air mixtures are explosive above flash point. Electrostatic discharges may be generated by flow or agitation resulting in ignition or explosion. EXTINGUISHING MEDIA: carbon dioxide, regular dry chemical. Large fires: Use regular foam or flood with fine water spray. FIREFIGHTING: Move container from fire area if it can be done without risk. For fires in cargo or storage area: Cool containers with water from unmanned hose holder or monitor nozzles until well after fire is out. If this is impossible then take the following precautions: Keep unnecessary people away, isolate hazard area and deny entry. Let the fire burn. Withdraw immediately in case of rising sound from venting safety device or any discoloration of tanks due to fire. For tank, rail car or tank truck: Stop leak if possible without personal risk. Let burn unless leak can be stopped immediately. For smaller tanks or cylinders, extinguish and isolate from other flammables. Evacuation radius: 800 meters (½ mile). Stop flow of gas.

    c. Toluene, 99% CAS #108-88-3
        From Fisher Scientific

        General Information: Containers can build up pressure if exposed to heat and/or fire. As in any fire, wear a self-contained breathing apparatus in pressure demand, MSHA/NIOSH (approved or equivalent), and full protective gear. Water runoff can cause environmental damage. Dike and collect water used to fight fire. Vapors may form an explosive mixture with air. Vapors can travel to a source of ignition and flash back. Use water spray to keep fire-exposed containers cool. Water may be ineffective. Material is lighter than water and a fire may spread by the use of water. Flammable Liquid. Vapors may be heavier than air. They can spread along the ground and collect in low or confined areas. May be ignited by heat, sparks, and flame. Containers may explode when heated. Extinguishing Media: Use water spray to cool fire-exposed containers. Water may be ineffective. Do NOT use straight streams of water. For large fires, use dry chemical, carbon dioxide, alcohol-resistant foam, or water spray. For small fires, use carbon dioxide, dry chemical, dry sand, or alcohol-resistant foam. Cool containers with flooding quantities of water until well after fire is out.

d. Hydrogen, CAS #1333-74-0
From MG Industries

FIRE AND EXPLOSION HAZARDS: Severe fire hazard. Severe explosion hazard. Vapor/air mixtures are explosive. Containers may rupture or explode if exposed to heat. Electrostatic discharges may be generated by flow or agitation resulting in ignition or explosion. EXTINGUISHING MEDIA: carbon dioxide, regular dry chemical. Large fires: Flood with fine water spray. FIREFIGHTING: Move container from fire area if it can be done without risk. Cool containers with water spray until well after the fire is out. Stay away from the ends of tanks. For fires in cargo or storage area: Cool containers with water from unmanned hose holder or monitor nozzles until well after fire is out. If this is impossible then take the following precautions: Keep unnecessary people away, isolate hazard area and deny entry. Let the fire burn. Withdraw immediately in case of rising sound from venting safety device or any discoloration of tanks due to fire. For tank, rail car or tank truck: Stop leak if possible without personal risk. Let burn unless leak can be stopped immediately. For smaller tanks or cylinders, extinguish and isolate from other flammables. Evacuation radius: 800 meters (½ mile). Do not attempt to extinguish fire unless flow of material can be stopped first. Flood with fine water spray. Cool containers with water spray until well after the fire is out. Apply water from a protected location or from a safe distance. Avoid inhalation of material or combustion by-products. Stay upwind and keep out of low areas. Evacuate if fire gets out of control or containers are directly exposed to fire. Evacuation radius: 500 meters (⅓ mile). Consider downwind evacuation if material is leaking. Stop flow of gas.

e. Regular unleaded gasoline, CAS #8006-61-9
From the Chevron Corporation

EXTINGUISHING MEDIA: Dry Chemical, CO_2, AFFF Foam, or alcohol resistant foam if >15% volume polar solvents (oxygenates). Reactivity 0. FIREFIGHTING INSTRUCTIONS: Use water spray to cool fire-exposed containers and to protect personnel. For fires involving this material, do not enter any enclosed or confined fire space without proper protective equipment, including self-contained breathing apparatus. COMBUSTION PRODUCTS: Normal combustion forms carbon dioxide and water vapor; incomplete combustion can produce carbon monoxide.

f. Ethyl ether, CAS #60-24-7
From Fisher Scientific

General Information: As in any fire, wear a self-contained breathing apparatus in pressure-demand, MSHA/NIOSH (approved or equivalent), and full protective gear. Vapors may form an explosive mixture with air. Vapors can travel to a source of ignition and flash back. Extremely flammable. Material will readily ignite at room temperature. Use water spray to keep fire-exposed containers cool. Water may be ineffective. Material is lighter than water and a fire may be spread by the use of water. Containers may explode in the heat of a fire. May form explosive peroxides. Vapors may be heavier than air. They can spread along the ground and collect in low or confined areas. Will be easily ignited by heat, sparks, or flame. May reignite after fire is extinguished. Extinguishing Media: For small fires, use dry chemical, carbon dioxide, water spray, or alcohol-resistant foam. Water may be ineffective. For large fires, use water spray, fog, or alcohol-resistant foam. Do NOT use straight streams of water. Cool containers with flooding quantities of water until well after fire is out.

g. Acetone, CAS #67-64-1
From Fisher Scientific

General Information: As in any fire, wear a self-contained breathing apparatus in pressure-demand, MSHA/NIOSH (approved or equivalent), and full protective gear. Vapors may form an explosive mixture with air. Vapors can travel to a source of ignition and flash back. Use water spray to keep fire-exposed containers cool. Extremely flammable liquid. Vapors may be heavier than air. They can spread along the ground and collect in low or confined areas. May be ignited by heat, sparks, and flame. Containers may explode when heated. Extinguishing Media: For small fires, use dry chemical, carbon dioxide, water spray or alcohol-resistant foam. Use water spray to cool fire-exposed containers. Water may be ineffective. Do NOT use straight streams of water. For large fires, use dry chemical, carbon dioxide, alcohol-resistant foam, or water spray. Cool containers with flooding quantities of water until well after fire is out.

h. Phosphorus (red), CAS #7723-14-0
From Fisher Scientific

General Information: As in any fire, wear a self-contained breathing apparatus in pressure-demand, MSHA/NIOSH (approved or equivalent), and full protective gear. Containers may explode in the heat of a fire. This material is readily combustible and may ignite if exposed to friction. Flammable solid. Vapors may be heavier than air. They can spread along the ground and collect in low or confined areas. May reignite after fire is extinguished. Extinguishing Media: For small fires, use dry chemical, sand, water spray, or foam. For large fires, use water spray, fog, or regular foam. Cool containers with flooding quantities of water until well after fire is out.

i. Acetylene, CAS #74-8602
From MG Industries

FIRE AND EXPLOSION HAZARDS: Severe explosion hazard. Vapor/air mixtures are explosive. Electrostatic discharges may be generated by flow or agitation resulting in ignition or explosion. EXTINGUISHING MEDIA: carbon dioxide, regular dry chemical. Large fires: Use regular foam or flood with fine water spray. FIREFIGHTING: Move container from fire area if it can be done without risk. For fires in cargo or storage area: Cool containers with water from unmanned hose holder or monitor nozzles until well after fire is out. If this is impossible then take the following precautions: Keep unnecessary people away, isolate hazard area and deny entry. Let the fire burn. Withdraw immediately in case of rising sound from venting safety device or any discoloration of tanks due to fire. For tank, rail car or tank truck: Stop leak if possible without personal risk. Let burn unless leak can be stopped immediately. For smaller tanks or cylinders, extinguish and isolate from other flammables. Evacuation radius: 800 meters (½ mile). Stop flow of gas.

j. Aluminum, rods, CAS #7429-90-5
From Fisher Scientific

General Information: As in any fire, wear a self-contained breathing apparatus in pressure-demand, MSHA/NIOSH (approved or equivalent), and full protective gear. During a fire, irritating and highly toxic gases may be generated by thermal decomposition or combustion. Extinguishing Media: Do NOT use water directly on fire. Use dry chemical to fight fire.

k. Ammonium hydroxide or aqueous ammonia, CAS #7732-18-5
From Fisher Scientific

General Information: As in any fire, wear a self-contained breathing apparatus in pressure-demand, MSHA/NIOSH (approved or equivalent), and full protective gear. During a fire, irritating and highly toxic gases may be generated by thermal decomposition or combustion. Use water spray to keep fire-exposed containers cool. Vapors may be heavier than air. They can spread along the ground and collect in low or confined areas. Contact with metals may evolve flammable hydrogen gas. Containers may explode when heated. Noncombustible, substance itself does not burn but may decompose upon heating to produce corrosive and/or toxic fumes. Extinguishing Media: For small fires, use water spray, dry chemical, carbon dioxide, or chemical foam. Do NOT get water inside containers. For large fires, use dry chemical, carbon dioxide, alcohol-resistant foam, or water spray. Cool containers with flooding quantities of water until well after fire is out.

2. How would you respond to the following situations?

   a. A small hot oil bath that is connected to an electrical heater catches fire. The surrounding area is free of any combustible materials.

   Notify co-workers of the fire while you unplug the heater and carefully place a cover on top of the oil bath. Use caution to avoid spilling or knocking over the flaming oil. Do not use a fire extinguisher. Carefully watch to be sure the fire is extinguished. If it is not, report it and follow local procedures for fire. It is important that the locations of breaker panel boxes be known and all breakers labeled. If the panel boxes are kept locked, a key must be readily available.

   b. An electronic pump catches fire during operation.

   Notify co-workers while you unplug the pump and/or shut off the electricity. Use a Class C fire extinguisher to put out this type of fire. If the fire is not quickly extinguished, follow local procedures for fire. It is important that the locations of breaker panel boxes be known and all breakers labeled. If the panel boxes are kept locked, a key must be readily available.

   c. A 5-gallon drum of acetone splits open and catches fire.

   Notify co-workers and immediately evacuate the room and surrounding areas. Sound a fire alarm and call for emergency assistance per local ordinances.

Section 7C: Eyewash Stations and Deluge Showers
Instructor Notes

In this section, students practice finding eyewash stations with their eyes closed and time each other getting to the deluge showers. Students also practice routine checks of eyewash stations and showers. The objective of this section is for students to understand when and how to use eyewash stations and deluge showers and how to check them for safe operation. Students learn that in a laboratory situation, a splash on the eye or skin requires immediate attention, with no time for panic or confusion about treatment. Proper treatment also requires fully functioning emergency equipment.

Student Pages

- Section 7C: Eyewash Stations and Deluge Showers—Background
- Section 7C: Eyewash Stations and Deluge Showers—Exercise

Possible Playout of the Section

Students read the Background and do the safety inspection Exercise. You may also wish to show students an emergency equipment training video called *Safety Net*, available from Haws Corporation ($7.95, 510/525-5801). Portions of the safety inspection may be assigned to different groups. Afterward, the groups report to the class on their results. The class discusses what to do if any of the laboratory equipment does not meet the standards described in the inspection instructions.

Setup Notes

During the Exercise, the students need access to a plumbed eyewash station and/or an eye-face wash station, and a deluge shower. The biggest challenge in this Exercise is preparing to collect the water generated during the tests. We suggest that you involve facility maintenance personnel in planning these tests. Be aware that if your equipment has not been tested in some time, the valves may stick.

Your eyewash and shower equipment may or may not be plumbed to a drain. If your eyewash station is not connected to a drain, you will need to create a method for collecting the water that is generated during the test so that students can measure the quantity and avoid creating a large spill. You may wish to solicit the help of maintenance personnel to accomplish this. (Note: If your eyewash station is connected to a drain, your students will not be able to measure the flow rate because they will have no practical way to collect the water.)

Plastic eyewash gauges, such as the Haws Model DT5 (shown in Figure 7C-3) are required to test the eyewash station. You may wish to photocopy Figure 7C-3 onto a sheet of overhead-projector acetate as a substitute for the actual item. The Exercise also requires a commercial or homemade deluge shower tester and a container large enough to hold the water generated during the test. In planning the size of container you need to collect the water, consider that some deluge shower installations provide greater volumes than the minimum 20 gallons per minute. (The shower test lasts only 15 seconds.) A deluge shower tester, which consists of a ring (to fit over the shower head) with a nylon funnel to channel water from the shower head to the bucket, is available from Flinn Scientific (Shower Safety Tester, cat. # SE1055, 800/452-1261, *www.flinnsci.com*) and Fisher Scientific (First Safety Shower Tester, cat. # 18-563, 800/766-7000, *www.fishersci.com*).

Section 7C: Eyewash Stations and Deluge Showers

Background

Note: This handout provides general recommendations for and information about emergency equipment. However, you should always follow the emergency plan established by your organization's chemical hygiene plan.

When a chemical splash occurs on your eyes or skin, you need an immediate source of clean, running water to flush the affected area. For this reason, eyewash stations and deluge showers are required in every academic or industrial laboratory. In the event of a chemical splash, you must know where this equipment is located, the equipment must be in good working order, and you must know exactly how the equipment works and for which types of emergencies you should use it. Keep in mind that wearing appropriate PPE, including protective eyewear equipment, is essential when using chemicals that could splash.

You must be able to get to eyewash and shower equipment very quickly in an emergency. Medical experts agree that in an eye injury involving chemicals, the first 10 seconds are the most critical in order to prevent severe or even permanent damage. OSHA regulations (29 CFR 1910.151) require that eyewashes "shall be provided within the work area for immediate emergency use." The American National Standards Institute (ANSI, "Emergency Eyewash and Shower Equipment Standard," Z358.1–1998) recommends that emergency stations be accessible within 10 seconds from a potential hazard. This time constraint usually means that an eyewash should be no more than 50 feet away from every laboratory work bench. Always keep aisles clear of furniture, open drawers, or any other obstacles so that a person moving to emergency equipment does not trip.

In the event of a chemical splash, someone should help the victim to the emergency equipment and summon professional help. Always seek medical attention in the event of a chemical splash because injuries or damage may not be readily apparent. As noted earlier, never work alone in a laboratory.

Treatment of a Person Contaminated by Spills or Splashes

The following procedures are recommended by *Prudent Practices in the Laboratory* for the treatment of spills or splashes covering small areas of skin:

- Immediately flush with flowing water for no less than 15 minutes.

- If no burn is visible, wash with tepid water and soap. Remove any jewelry so that any residual materials can be cleaned.

- Check the MSDS to see whether any delayed effects should be expected.

- Seek medical attention for even minor chemical burns.

- Do not use any creams, lotions, or salves without medical approval.

- Do not attempt to neutralize the splashed chemical.

When working with a chemical that poses extreme hazards, make sure you know about special emergency procedures that can help prevent serious injury. For example, new emergency first aid treatments for exposure to hydrofluoric acid (ranked as one of the compounds most hazardous to human health) are reducing the severe (even fatal) consequences of HF contact by skin, eye, or breathing. (For details see, "First aid for a unique acid, HF: A sequel," *Chemical Health & Safety*, January/February 2000, 18–23.)

These steps for the treatment of spills on clothes are adapted from *Prudent Practices:*

- Do not attempt to wipe the clothes.

- Take care not to spread the chemical on the skin or, especially, into the eyes.

- Quickly remove all contaminated clothing, shoes, and jewelry while using the deluge shower. Be careful while removing clothing; the individual could suffer from acid burns if contaminated clothing comes into contact with exposed skin. Do not waste time because of modesty; seconds count.

- Use caution when removing pullover shirts or sweaters to prevent contamination of the eyes; it may be better to cut the garments off.

- Wash head before removing goggles.

- Immediately flood the affected body area with tepid water for at least 15 minutes.

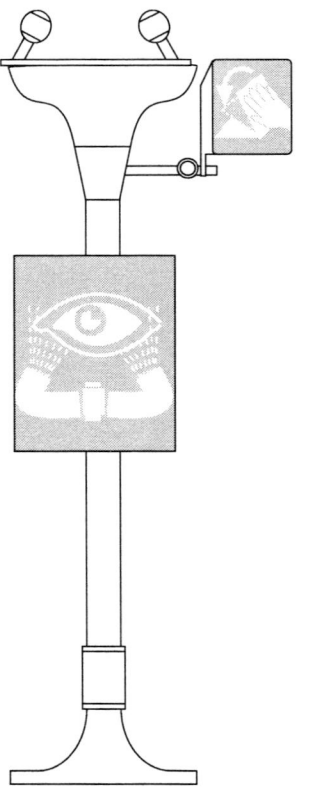

- Get medical attention as soon as possible.

- Discard contaminated clothes or have them laundered separately from other clothing.

Prudent Practices recommends these steps for the treatment of splashes in eyes:

- Immediately flush with tepid, potable water from a gently flowing source for at least 15 minutes. Use an eyewash if available. If not, place the injured person on his or her back and pour water gently into the eyes for at least 15 minutes.

- Hold the individual's eyelid away from the eyeball, and instruct him or her to move the eye up and down and sideways to thoroughly wash behind the eyelids.

- Follow washing with prompt treatment by a medical professional who is acquainted with chemical injuries.

Figure 7C-1:
An eyewash station

Common Types of Eyewashes and Deluge Showers

Eyewash stations, such as the example shown in Figure 7C-1, are designed to irrigate and flush the eyes. When turned on, the stations gently but continuously spray water toward both of the user's eyes. Many models are available, and you need to be familiar with the operation of the eyewash stations in your laboratory. If you splash a chemical in your eyes, flush your eyes out for at least 15 minutes at a minimum of approximately 0.4 gallons per minute, or approximately 3 gallons per minute if it is an eye/face wash. Eye/face wash stations are similar to eyewash fountains except that the eyes and face are flushed simultaneously. This type of equipment might provide more help than an eyewash station in an emergency if a laboratory worker splashed chemicals on his or her entire face, not just in his or her eyes.

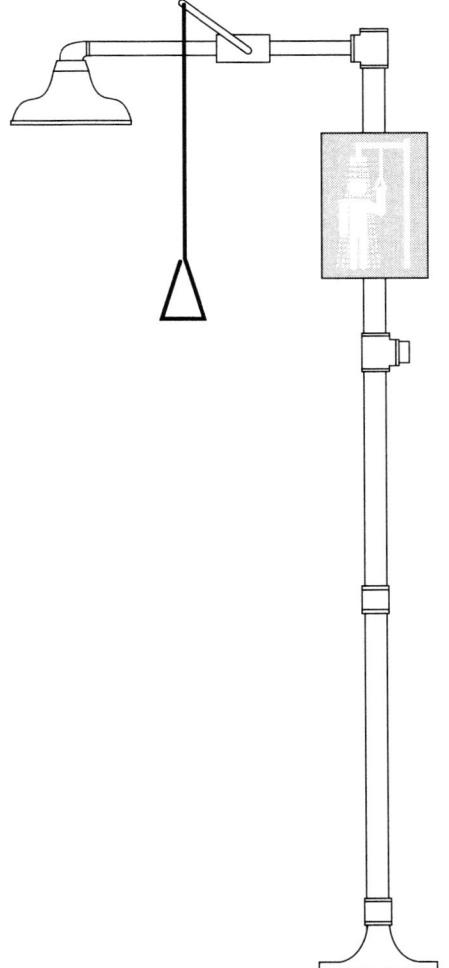

Figure 7C-2: An emergency shower

Emergency showers, such as the example shown in Figure 7C-2, also referred to as deluge or drench showers, pour a minimum of 20 gallons of water per minute on a victim. This amount of water is what a regular-sized bathtub holds. These showers are useful not only for chemical splashes but also for victims whose clothing is on fire. However, a person whose clothing is on fire should never run across a room to reach a deluge shower. A shower should be used to douse a person's clothes only if the fire victim is very close by. To reduce the amount of damage to a person's skin, the person should remove affected clothing while using a safety shower, especially in the event of a corrosive liquid spill. Some showers have a curtain to provide privacy. If not, co-workers can hold up a fire blanket. Some institutions keep disposable jump suits on hand in case someone's clothes become contaminated.

Deluge showers should not be used for eye emergencies because of the amount of water that is released. Also, these showers should never be located near electrical equipment because so much water is released so quickly. Currently, safety experts are debating about whether showers should have drainage systems and, if so, what type is best.

Combination units are simply a combination of a deluge shower and eyewash (or eye/face wash) station. These units are designed so that victims can use both faucets simultaneously or just one or the other.

Make Sure Your Equipment Is Working Properly

Proper installation and maintenance of safety equipment are vital for a safe laboratory. Because every second counts when getting to this emergency equipment, everyone in the laboratory should consider it is his or her own responsibility to ensure that the eyewash or deluge shower is clearly labeled and accessible.

The ANSI Standard requires a "stay-open valve" on eyewash stations and deluge showers so victims can use both hands to keep their eyelids open while flushing; to remove affected clothing if necessary in a deluge shower; and to encourage flushing for at least 15 minutes. However, students must realize that eyewash stations can have one of several types of valves and they should become familiar with the type(s) in their laboratory. It is a good idea to have the eyewash and deluge shower equipped with an alarm that sounds when water is flowing through the system.

According to Appendix A of the OSHA Laboratory Standard (29 CFR 1910.1450), eyewash fountains "should be inspected at intervals of not less than three months," and other safety equipment "should be inspected regularly" (for example, every 3–6 months). ANSI recommends that emergency equipment should be tested weekly. Shower test kits are available through most emergency equipment suppliers. Although not mandated, a weekly test is recommended to verify that the water line to the unit has not been inadvertently closed, to verify that the equipment is working properly, to flush the lines of any particles or debris that may harm someone trying to use an eyewash or deluge shower, and to discourage any microbial growth. According to a bulletin issued by the Assistant Secretary for Environment, Safety, and Health of the US Department of Energy, a small infectious amoeba, *Acanthamoeba,* has been found in numerous portable and stationary eyewash stations at a Department of Energy facility. This amoeba is capable of causing serious eye infections—some victims have lost the infected eye. The National Safety Council recommends a 3-minute flush of plumbed eyewash installations weekly to reduce the threat of such infections. It is recommended that a record be kept of such tests.

Chemical and/or safety supply companies sell gauges and kits to test eyewash and shower equipment. An eyewash gauge (such as the Haws Model DT5 shown in Figure 7C-3) will tell you whether the stream of water from the eyewash station will hit a chemical splash victim in the right spot. When you place the gauge over the stream of the eyewash, the water should cover the areas designated on the gauge.

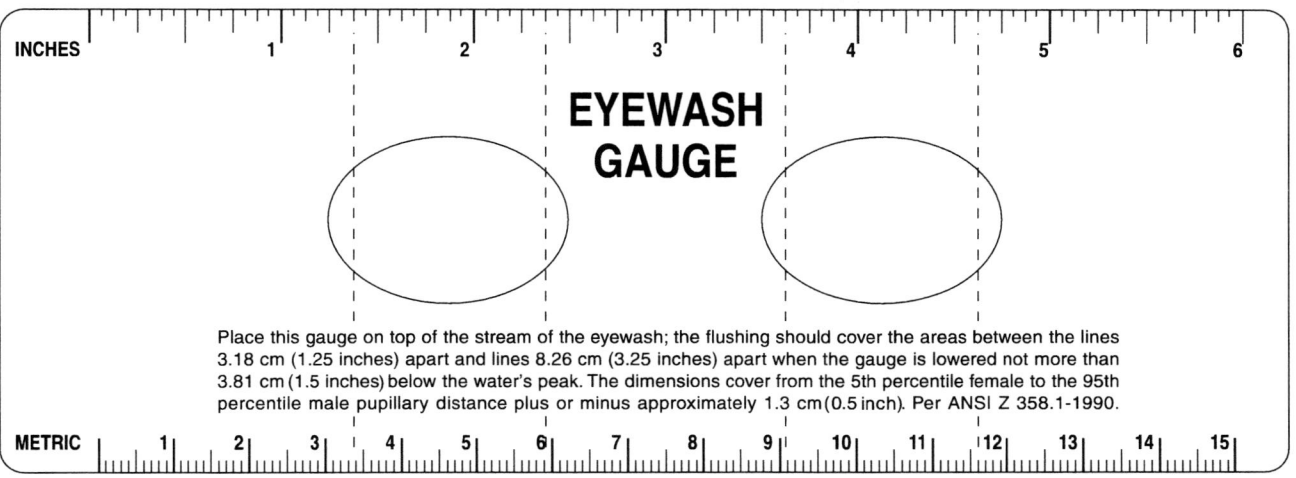

(actual size)

Figure 7C-3: An eyewash gauge

A deluge shower test kit allows you to determine whether the shower is working properly without the potential mess, particularly if the area is not equipped with a drain. These kits attach to the head of a deluge shower and direct the water flow through a waterproof fabric funnel into a bucket.

Section 7C: Eyewash Stations and Deluge Showers
Exercise

In this Exercise, you will practice finding eyewash stations with your eyes closed and time each other getting to the deluge showers. You will learn when and how to use eyewash stations and deluge showers and how to check them for safe operation.

Working in assigned groups, evaluate your eyewashes, combination eye-face washes, and/or deluge showers according to the ANSI standards in the tables on the following two pages. The exercise requires an eyewash gauge (provided by your instructor) to test for the proper orientation of the flow stream in the eyewash station, a tape measure to verify that the eyewash station and shower heads are the proper distance from the floor and walls, and a shower tester. You will also need containers to collect and measure the flow from the deluge shower and from the eyewash station (if they are not plumbed to a drain). Once you have the necessary materials, fill in the tables.

Plumbed Eyewash Stations		
ANSI Standards Z358.1-1998		Observation/Measurement of Lab Equipment
Eye-Face Washes	Eye-Only Washes	
Heads positioned 33–45 inches from floor		
Heads positioned 6 inches from wall or nearest obstruction		
Equipment has large heads to cover both eyes and face or regular size eyewash heads plus a face spray ring	Not applicable	
Flow rate will be a minimum of 3 gallons per minute (gpm) for 15 minutes ***To Test**: Test for 30 seconds; collect water in a bucket; measure.	Flow rate will be a minimum of 0.4 gallons per minute (gpm) for 15 minutes ***To Test**: Test for 1 minute; collect water in a bucket; measure.	
Valves activate in 1 second or less		
Valves stay open with no use of hands		
Installation is 10 seconds from the hazard **To Test**: Choose an "eye buddy" to assist you in the process and prevent injury. Stand at the center of the lab area where potentially dangerous chemicals are used. Blindfold your partner (this simulates an actual emergency more accurately) and have him or her find the eye/facewash station in 10 seconds or less with your assistance so the "victim" doesn't run into anything or get hurt in the process.		
Location in a well-lit area, identified by sign		
Other Considerations		
Water flow positioned according to eyewash gauge		
Accessible from three directions		
Safe distance from anything electrical		
Free from airborne dust and debris of any kind		
*You will only be able to test flow rates of eye-face and eye-only washes if the equipment is **not** plumbed to a drain.		

Plumbed Showers	
ANSI Standards Z358.1–1998	Observations/Measurements of Lab Equipment
Heads positioned 82–96 inches from floor	
At 60 inches above the floor, spray pattern will have a minimum diameter of 20 inches	
Flow rate will be a minimum of 20 gallons per minute (gpm) as long as the appropriate flow pattern is maintained at 30 pounds per square inch (psi) **To Test:** Use the commercial shower test kit or homemade equivalent to test for 15 seconds. Collect water in a container.	
Valves activate in 1 second or less	
Valves stay open after activation with no use of hands	
Installation is 10 seconds from the hazard **To Test:** Choose a "buddy" to assist you in the process and prevent injury. Stand at the center of the lab station where potentially hazardous chemicals are used. Blindfold your partner (this simulates an actual emergency more accurately) and have him or her find the shower in 10 seconds or less with your assistance so the "victim" doesn't run into anything or get hurt in the process.	
Location in a well-lit area, identified by sign	
Other Considerations	
Accessible from three directions	
Safe distance from anything electrical	
Free from clutter and debris of any kind	

Section 7D: Responding to Spills
Instructor Notes

The objective of the section is to prepare students to use spill kits appropriately in response to spills of various hazardous materials. Students practice spill-response techniques using vinegar as a substitute for more corrosive materials.

Student Pages

- Section 7D: Responding to Spills—Background
- Section 7D: Responding to Spills—Lab: Spill Response

Possible Playout of the Section

Students read the Background and complete the Lab.

Safety

The labs in this book do not routinely include special safety precautions for the chemicals or equipment that might be used. As the course instructor, it is expected that you will provide students with access to SOPs, MSDSs, and other resources they need to safely work in the laboratory while meeting all regulatory requirements. Before doing any of the labs from this book or from other sources, we recommend that you regularly review special handling issues with students, allow time for questions, and then assess student understanding of these issues.

Setup Notes

Materials per group
- bottle of vinegar
- cat litter
- baking soda (sodium bicarbonate)
- plastic broom
- pH paper
- resealable storage bags

Section 7D: Responding to Spills

Background

Note: This handout provides general recommendations for and information about emergency equipment. However, you should always follow the emergency plan established by your organization's chemical hygiene plan.

While the first and most important step in spill control is spill prevention, you must be prepared to respond to a spill in a laboratory in which you are working. Chemical spills in laboratories may be easily contained and cleaned up, or they may be much more extensive and pose a serious threat to everyone in the laboratory. The physical size of a chemical spill may not be a true indication of its severity and whether you should evacuate.

Before You Clean Up a Chemical Spill

Every laboratory worker must be prepared to respond to a chemical spill. In order to clean up a chemical spill, you must take several important factors into consideration. First, you must correctly identify the spilled material. If you spilled the chemical while working with it, you should already have read the MSDS and know the hazards associated with the chemical and be using the appropriate PPE. Container labels should also tell you some of what you need to know.

Next, you must eliminate any additional hazards. For example, if the spilled chemical is flammable, make sure all flames and ignition sources are extinguished. Communication and teamwork among laboratory co-workers are important. Unless you are working alone in a laboratory—which is a very bad idea—you and your co-workers can share the necessary steps to quickly contain a spill while minimizing the danger to any individuals. If you see a spill occur, make sure everyone knows where it is and what it is immediately.

Be very careful about spills on tile floors. Most tile floors become very slippery when wet. Acids and especially bases can sometimes be even more slippery and pose an extreme hazard, that of slipping and falling into the acid or base spill. Never walk on a spilled material.

Containing and Cleaning Up a Chemical Spill

All laboratories should have sufficient spill control materials to handle a spill from the largest container used in the laboratory. Spill control kits tailored to deal with different types of spills should be available. Spill control kits will typically contain the following items:

- spill control pillows for absorbing solvents, caustic alkalis, and acids (but not hydrofluoric acid);

- inert absorbents such as vermiculite, clay, and cat litter (paper towels are not recommended as an absorbent material for oxidizing agents);

Building Student Safety Habits for the Workplace

- absorbents that control vapors, such as Solusorb®, an activated charcoal product manufactured by J.T. Baker (Solusorb, cat. # 4458-05, 800/582-2537);

- neutralizing agents (such as sodium bicarbonate) for acid spills;

- neutralizing agents (such as citric acid) for alkali spills;

- plastic scoops, pails, and dust pans; and

- appropriate PPE, warning signs, and barricade tapes.

If nothing else is available, a wool fire blanket will work too. Be sure to dispose of the blanket as waste after the spill is controlled.

Federal law prohibits cleaning up other people's hazardous spills without proper training (29 CFR 1910.1200, Hazardous Waste Operations Emergency Response). *Prudent Practices* states that "Specific procedures for cleaning up spills vary depending on the location of the accident, the amount and physical properties of the spilled material, the degree and type of toxicity, and the training of the personnel involved." Outlined here are some general guidelines from safety consultant Rudy Gerlach for handling **small** spills. However, always follow the procedures outlined by your chemical hygiene plan and never attempt to clean up a spill without proper training.

- Inorganic acids and bases: Wear appropriate PPE. Contain and neutralize the spill by pouring a neutralizing material such as sodium bisulfate (for bases) or sodium carbonate or bicarbonate (for acids) around and over the liquid. Mix the neutralizer and spilled chemical with a plastic broom to maximize contact and neutralization. Use an appropriate device (such as a plastic dust pan) to scoop waste into an appropriate container for disposal.

- Flammable solvents: Wear appropriate PPE. Contain the spill and its vapors by pouring a vapor-containing absorbent such as Solusorb around and over the spill. Mix the absorbent and spilled chemical with a plastic broom. Scoop into appropriate container for disposal using nonsparking tools.

If you are unsure about how to respond to a spill in your laboratory, it is better just to evacuate the area and call emergency personnel rather than risk making the spill spread or causing injury to yourself or to others. Practicing your response will make it easier to respond quickly and correctly in the event of a real spill and will also allow you to become familiar with where the spill control materials are stored in your laboratory.

Be sure you know how to properly dispose of the chemical once you have cleaned it up. The MSDS may provide some information, but local and state regulations differ. Deposit all waste in appropriately labeled receptacles and follow all other waste disposal procedures in your organization's chemical hygiene plan. Indiscriminate disposal of waste by adding it to regular trash for landfill burial is never acceptable.

Common Types of Laboratory Spills

The following table describes some spills common in a laboratory setting.

Common Laboratory Spills		
Spill Characteristics	Examples	How to Handle
Non-flammable, non-volatile, and low-toxicity liquids	aqueous solutions of inorganic salts	Use an inert absorbent, and dispose of the material properly.
Non-toxic, non-flammable solids	sodium chloride, potassium sulfate	Restrict unprotected personnel from the area. Sweep up the material, place it in a sealed bag or container, and dispose of the waste. Ventilate the area, and wash the surface after material pickup is complete.
Corrosive materials (acidic)	hydrochloric acid, sulfuric acid	Restrict unprotected personnel from the area. Remove all ignition sources, and ventilate the area. Contain the spill with sand and absorbent material, neutralize the spill with sodium bicarbonate, and deposit the waste in a sealed bag or container.
Caustic materials (basic)	sodium hydroxide solution, aqueous ammonia	Restrict unprotected personnel from the area. Remove all ignition sources, and ventilate the area. Contain the spill with sand and absorbent material, neutralize the spill with citric acid or sodium bicarbonate, and deposit the waste in a sealed bag or container.
Flammable solvent of relatively low toxicity	petroleum ether, pentane, diethyl ether	Extinguish all flames in the laboratory, turn off any equipment that may produce sparks, and ventilate area. Use a spill absorbent that controls vapors. Seal these in containers and dispose of them properly. Use non-sparking tools in the cleanup. Be careful of excessive vapors!
Flammable, combustible solids	charcoal, butyl stearate	Restrict unprotected personnel from the area. Ventilate the area. Remove all ignition sources and water. Sweep up the material, place it in a sealed bag or container, and dispose of the waste. Wash the spill site after material pickup is complete.
Highly volatile, toxic materials	concentrated aqueous ammonia, butyric acid	Do not try to clean up these spills. Contact the appropriate safety and industrial hygiene personnel.
Elemental mercury	mercury thermometer or barometer break	Metallic mercury is highly toxic by skin absorption and inhalation of vapor. Isolate the spill area and provide adequate ventilation. Wear protective gloves and use a scraper or piece of cardboard to pool the droplets together, depending on the quantity of mercury. Clean up the pool with a special vacuum pump (do not use a regular vacuum cleaner), wet toweling, adhesive tape, or a commercial mercury spill control kit. Use an appropriate decontamination kit to clean the exposed work surfaces. Test for residual mercury after cleanup. The vacuum pump will need to be decontaminated.

Section 7D: Responding to Spills
Lab: Spill Response

This Lab will give you practice in responding to a spill of a chemical.

Safety

In a laboratory setting, you are ultimately responsible for your own safety and for the safety of those around you. In order to more closely resemble an actual workplace, specific safety information is not routinely provided to you for the labs in this book. As discussed throughout the book, many resources are available to assist you in obtaining the information you need. It is your responsibility to specifically follow your institution's standard operating procedures (SOPs) and all local, state, and national guidelines on safe handling and storage of all chemicals and equipment you may use in the labs. This includes determining and using the appropriate personal protective equipment (e.g., goggles, gloves, apron). If you are at any time unsure about an SOP or other regulation, check with your instructor.

Procedure

1. Put on the appropriate PPE. Create a small spill by pouring about 1 cup vinegar on the laboratory bench or floor.

2. Respond to the spill appropriately using the materials provided by your instructor.

3. Hypothesize how cleaning spills of each of the chemicals listed below would be different than cleaning the vinegar spill. **Do not actually spill these chemicals.** Consider issues such as reaction rate, CO_2 generation, splatter, use of different neutralizers, or whether to use a neutralizer at all.
   - concentrated H_2SO_4
   - 1 M HCl
   - solid NaOH
   - 6 M HNO_3
   - concentrated NH_3 (aq)
   - toluene
   - hexane
   - solid phosphoric acid
   - elemental Hg

   See the Common Laboratory Spills table for responses to these spills.

4. Check your chemical hygiene plan to see if it includes procedures for responding to spills of the chemicals listed in step 3. If not, work as a class to establish a plan for handling these types of spills.

Chapter 8
Safety Planning and Standard Operating Procedures

I was performing an organic synthesis in my undergraduate lab program using toluene as a solvent. After observing the proper precautions to heat the apparatus, I reached a point in the synthesis at which additional chemicals were required. This step required cooling the apparatus and turning off the Bunsen burner. However, I chose instead to move two or three feet away and open the round-bottomed flask to add chemicals without cooling or turning off the Bunsen burner. Toluene is volatile at room temperature but even more so at the increased temperature. The vapors spread across the lab bench and ignited, reaching the flask in my hand. Fortunately only first-degree burns and embarrassment resulted. The fire was quickly put out with an asbestos pad covering the flask.

Learning by Accident, Vol. 2, page 48

Chapter 8 Objectives

Pertinent Voluntary Industry Standards

- *VIS L2 Standards*—Recognize that each company has policies and safety plans that include evacuation procedures, emergency numbers, rules, and practices.

- *VIS L2 Standards*—Be aware of and follow federal, state, and local legislation pertaining to environmental, health, and safety regulations; identify their responsibilities under these regulations.

- *VIS (L2.02):01*—Identify the responsibilities of the technician for applying regulatory guidelines in a variety of typical laboratory situations.

- *VIS L8 Tasks*—Familiarizing oneself with reaction characteristics before performing syntheses, including the nature of the reaction (kinetics, equilibrium, exothermic/endothermic, etc.).

- *VIS L8 Tasks*—Determining all safety and handling aspects of the work to be conducted (for example, consider side reactions).

- *VIS L8 Tasks*—Disassembling, cleaning, and returning equipment, glassware, and other materials used to proper stored conditions.

- *VIS (L8.05):01*—Locate and use pertinent chemical reference materials.

Objectives of the Chapter

Students will learn

- about OSHA, NIOSH, ANSI, and the Voluntary Industry Standards;

- how to do a simplified risk analysis;

- how to analyze an SOP for safety procedures;

- how to write an SOP for a routine laboratory procedure; and

- how to plan and conduct safety meetings and audits.

Why Your Students Need to Understand Safety Planning and Standard Operating Procedures

Industry demand and the safety of your students during their education and in their eventual workplaces make learning about safety planning and SOPs an important part of a safety curriculum.

Students need to be aware of government regulations and regulatory agencies because part of their jobs as science professionals may involve dealing directly with such regulations and agencies. Students should be able to find regulations, research and understand regulations as they change, and incorporate those regulations into their industry laboratory procedures.

Safety planning is part of any good industrial or academic laboratory workplace. In industry, laboratory professionals are expected to implement and comply with safety procedures. Chemists, chemical technicians, and other laboratory professionals should be actively involved in safety planning since they are working in a high-risk environment on a day-to-day basis. Students, therefore, need to be prepared for the safety culture of industry before they begin working in it.

SOPs are a part of safety planning and can prevent accidents and injuries if individuals make an effort to follow them scrupulously. Since chemical technicians and other laboratory professionals are routinely responsible for writing SOPs, they need to have opportunities to practice this skill and learn that following SOPs can enhance workplace safety.

Overview of the Chapter

Chapter 8: Safety Planning and Standard Operating Procedures contains four sections. General background information and complete instructions for the exercises are provided. The background pages are intended only to provide basic background information on safety planning and SOPs for your students. We recommend that you extend this background information by having students find and read the pertinent sections of the books in the safety reference library (described in Chapter 1). You may wish to have students briefly report on the information they find during the monthly safety meetings.

Section 8A: Getting to Know Safety Planning and SOPs

This section introduces students to safety planning and SOPs. To initiate interest in the topic, students read and discuss Laboratory Incident Reports concerning accidents that resulted from inadequate safety planning or lack of safety information in SOPs.

Section 8B: Risk Assessment

As preparation for evaluating an actual SOP for safety, students perform a risk assessment on using a common piece of laboratory equipment—a Bunsen burner. The objective of the Exercise is for students to learn that step-by-step written instructions not only allow the same task to be done consistently by many different people, but can also serve as an outline for brainstorming potential hazards associated with a task.

Section 8C: Evaluating an SOP

The objective of this section is for students to learn about SOPs. They read and analyze an existing SOP to determine whether it includes enough safety procedures, applying all the other specific procedures they have learned from this book.

Section 8D: Safety Meetings and Audits

This section is designed to help students understand the role of safety meetings and audits in laboratory safety planning.

References for Chapter 8

About NIOSH. www.cdc.gov/niosh/about.html (accessed Sept 25, 1998).

American National Standards Institute. www.ansi.org/default_js.htm (accessed Feb 5, 2000).

Casey, S. *Set Phasers on Stun and Other True Tales of Design, Technology, and Human Error;* Aegean: Santa Barbara, CA, 1993.

Guidelines for Hazard Evaluation Procedures, 2nd Edition with Worked Examples; American Institute of Chemical Engineers; Center for Chemical Process Safety: New York, 1992.

Guidelines for Incorporating Safety and Health into Engineering Curricula, Vol. 1: Laboratory Safety; Joint Council for Health, Safety, and Environmental Education of Professionals: Savoy, IL, 1994.

Hazard Communication Standard. *Code of Federal Regulations.* 29 CFR 1910.1200, 1999.

Hofstader, R.; Chapman, K. *Foundations for Excellence in the Chemical Process Industries: Voluntary Industry Standards for Chemical Process Industries Technical Workers;* American Chemical Society: Washington, DC, 1997.

Improving Safety in the Chemical Laboratory: A Practical Guide; Young, J.A., Ed.; John Wiley & Sons: New York, 1991.

Joint Council for Health, Safety, and Environmental Education of Professionals, *Guidelines for Incorporating Safety and Health into Engineering Curricula, Vol.1, Laboratory Safety,* 1994.

Kinsley, W.K. "Helping to achieve a high level of proficiency, production, and safety." *Chemical Health & Safety,* 1988, 5, 28–31.

Laboratory Standard. *Code of Federal Regulations.* 29 CFR 1910.1450, 1999.

Learning by Accident, Vol. 2; Mojtabai, F.; Kaufman, J., Eds.; Laboratory Safety Institute: Natick, MA, 2000.

Thomen, J.R. *Excellence in Safety Leadership;* John Wiley & Sons: New York, 1991.

U.S. Department of Energy. Lessons Learned Information Services Homepage. tis.eh.doe.gov/ll/ (accessed Feb 25, 2000).

U.S. Department of Labor, *Process Safety Management of Highly Hazardous Chemicals,* Fact Sheet No. OSHA 92-45.

Section 8A: Getting to Know Safety Planning and SOPs
Instructor Notes

This section introduces students to safety planning and SOPs. To initiate interest in the topic, students read and discuss Laboratory Incident Reports concerning accidents that resulted from inadequate safety planning or lack of safety information in SOPs.

Student Pages

- Section 8A: Getting to Know Safety Planning and SOPs—Laboratory Incident Reports
- Section 8A: Getting to Know Safety Planning and SOPs—Background

Possible Playout of the Section

Students read the Laboratory Incident Reports and Background and discuss the following questions:

1. What are the similarities and differences between the Laboratory Incident Reports?

2. What conclusions can you draw about safety from these incidents?

3. How could each of these incidents have been prevented?

Section 8A: Getting to Know Safety Planning and SOPs
Laboratory Incident Reports

I was performing an organic synthesis in my undergraduate lab program using toluene as a solvent. After observing the proper precautions to heat the apparatus, I reached a point in the synthesis at which additional chemicals were required. This step required cooling the apparatus and turning off the Bunsen burner. However, I chose instead to move two or three feet away and open the round-bottomed flask to add chemicals without cooling or turning off the Bunsen burner. Toluene is volatile at room temperature but even more so at the increased temperature. The vapors spread across the lab bench and ignited, reaching the flask in my hand. Fortunately only first-degree burns and embarrassment resulted. The fire was quickly put out with an asbestos pad covering the flask.

Learning by Accident, Vol. 2, page 48

A laboratory assistant in a large university was cleaning storage bottles with diluted acid solution. Proper techniques had been emphasized except for one minor detail. No one had told the student to not place a cap on the container being cleaned. After cleaning 25 or 30 bottles, the assistant temporarily placed a cap on one of the bottles. The bottle exploded. . . . The top one-third of the broken bottle is now secured above the sink in the prep room as a reminder.

Learning by Accident, Vol. 2, page 64

On June 12, 1995, a graduate student working at the Health Research Laboratory at Los Alamos National Laboratory complained of common viral symptoms to the occupational medicine group. The student wondered whether the symptoms were caused by chemicals used in the laboratory. Investigation by the physician assistant and the graduate student revealed that on June 8, a post-doctorate employee using a funnel to transfer hazardous liquid chemical waste from a peptide synthesizer instrument disposal container to a waste bottle spilled 200–500 mL of the liquid on the floor. He inappropriately used paper towels to absorb the spill and a custodial mop to clean up the residue, instead of using a nearby spill kit designed for the situation. He threw the paper towels into administrative trash, which resulted in the hazardous waste being transported to the Los Alamos County landfill. The employee and a graduate student who had observed the spill left without informing facility managers about the spill. Thus, other individuals were needlessly exposed to the hazardous waste. As a result of this incident, the Life Sciences manager suspended work with the peptide synthesizer instrument until a standard operating procedure for using the instrument was developed, a user's list for the instrument established, and an experiment review procedure established.

DOE, Lessons Learned Information Services Home Page

On the evening of January 3, 1961, the worst nuclear accident in U.S. history occurred at the Atomic Energy Commission's National Reactor Testing Station near Idaho Falls, Idaho. The S-1 test reactor had been running for nearly two years with few problems other than an occasional stuck control rod. The reactor was a prototype for compact nuclear power plants that the Army hoped to use at remote military bases in the Arctic.

The reactor used nine control rods to keep the uranium in the reactor's core from reaching a "critical mass" that could result in an uncontrolled nuclear chain reaction. The rods extended upward from the core through a surrounding pressure vessel. At the top of the pressure vessel was the reactor, which contained nine bell-shaped drive housings to cover the rods and the motorized drive mechanisms needed to move them. Earlier that day, workers had disassembled the drive housings and mechanisms to install special test equipment into the core. Three young servicemen, working the second shift, were instructed to reassemble the drive mechanisms and drive housings so that the reactor could be re-pressurized with coolant water and running by the next day.

The Army had a standard operating procedure (SOP) for disassembling the drive housing that required the control rod to be raised to a height of no more than 4 inches above the top of the reactor's pressure vessel and secured with a C-clamp. For reassembly, the SOP merely stated that the procedure was the reverse of disassembly. In fact, the men quickly discovered that reassembly required raising the control rod to a height of at least 7 inches. Then, to remove the clamp, the men needed to lift the rod a fraction of an inch higher. As they attempted to do this, the rod became stuck. Two of the men combined their strength and pulled sharply upward on the rod. With no mechanical stops to keep the rod from rising too high, the rod bolted upward before the men could release their grip. With the rod partially extracted, there was now a critical mass of uranium within the core. In just thousandths of a second, the core temperature shot to 3,740°F, melting half the fuel elements in the core. The standing water in the "shutdown" core instantly vaporized, rupturing the pressure vessel, and releasing fragments of the core and other intensely radioactive debris throughout the containment building.

By the time rescue workers arrived, radiation levels in the containment building had reached 1,000 rem per hour—many times the lethal dose. Though wearing heavy radiation suits, rescuers could work no more than 3-minute shifts to search for and retrieve the victims. One of the servicemen was still alive but was unrecognizable due to his injuries. He died about an hour later. The other two were dead by the time rescue workers arrived. All three men were buried in lead coffins after repeated but failed attempts at decontamination.

Decontamination and disassembly of the reactor took another two years. In the subsequent investigation, the designers revealed that they knew that removal of the reactors' central control rod was theoretically sufficient to cause a runaway chain reaction. The designers simply did not anticipate removal of a control rod during operation. Blame for the accident was placed on the three servicemen who had removed the control rod beyond the limit specified in the maintenance SOP. Still, the reactor housing had no markings or mechanical stops to show the level of the rod. The workers had to estimate the distance. The SOP had no warning regarding the consequences of raising the rod beyond the limit specified.

Summarized from *Set Phasers on Stun* (pages 117–132)

Section 8A: Getting to Know Safety Planning and SOPs
Background

Note: This section provides general recommendations for and information about safety planning and SOPs. However, you should always follow the procedures established by your organization's chemical hygiene plan.

"Is there anything else we failed to anticipate in our safety plan?"

A safe working environment begins with safety planning. Effective safety planning must draw upon a detailed knowledge of prudent laboratory practices, regulatory requirements, and risk evaluation techniques. The written products of safety planning include a chemical hygiene plan (CHP) and standard operating procedures (SOP). Safety meetings and laboratory safety audits help maintain the safe behaviors and environments that a CHP and SOPs outline. These are plans of good practice. If followed, they will prevent employees from being overexposed to hazardous chemicals.

Regulatory and Standards Oversight Groups

This section provides an overview of the different agencies and regulations that are involved in workplace safety. The CHP and SOPs you will follow in the workplace must comply with these regulations, and employers expect their newly hired laboratory employees to be familiar with them. Much of this regulatory information has been discussed in other chapters; this chapter provides a summary.

Occupational Safety and Health

In 1970, President Richard Nixon signed the Williams-Steiger Occupational Safety and Health Act (OSH Act) into law. The law created the Occupational Safety and Health Administration (OSHA). OSHA is a regulatory body that creates and enforces workplace safety regulations. (You can visit OSHA on the Web at *www.osha.gov.*) OSHA oversees compliance with the Hazard Communication Standard and the Laboratory Standard. These regulations are published in the Federal Register and in Title 29 of the Code of Federal Regulations (CFR) Part 1910 (General Industry). They contain 26 subparts (A–Z) and nearly 1,500 separate sections. OSHA regulations apply to private-sector employers in all states and territories. Public-sector employees are covered only if the state or territory has a "State Plan" for enforcing workplace safety and health regulations or other applicable state law. Compliance with 29 CFR 1910 is overseen by OSHA. The Occupational Safety and Health Review Committee hears appeals to citations and fines. Appeals can be filed by any employer found in violation of OSHA standards. Employees can keep up-to-date with OSHA regulations through the agency's homepage at *www.osha.gov.* The site contains timely news releases, congressional testimony, interpretations, directives, and special reports.

The National Institute of Occupational Safety and Health (NIOSH) conducts health and safety research and proposes regulations. (You can visit NIOSH on the Web at *www.cdc.gov/niosh/homepage.html.*) NIOSH is not a part of OSHA. It is a division of the Department of Health and Human Services and conducts research on the causes of work-related injuries and illnesses. NIOSH emphasizes prevention and reduction of workplace dangers by determining effective ways to protect workers.

The American National Standards Institute (ANSI) develops non-regulatory consensus standards with input from the specific industries for which the standards are developed. These standards are adopted voluntarily by various organizations, companies, and industries. While these standards lack the regulatory force of OSHA regulations, they are taken seriously by most employers and are therefore important to you. In some cases, OSHA adopts an ANSI standard, and then the voluntary standard takes on the force of a regulation. ANSI's website at *www.ansi.org* is designed to provide access to timely information on the latest national and international standards-related activities. (Membership is required to access some information.)

Hazardous Chemicals in the Workplace

With regard to hazardous chemicals in the workplace, three sections of the OSHA regulations are particularly important: the Hazard Communication Standard (29 CFR 1910.1200), the Laboratory Standard (29 CFR 1910.1450), and the Process Safety Management (PSM) Standard for highly hazardous chemicals (29 CFR 1910.119). You can find the Hazard Communication Standard on the Web at *www.osha-sk.gov/OshStd_data/1910_1200.html.* The Laboratory Standard is at *www.osha-sk.gov/OshStd_data/1910_1450.html.* Further information on the PSM Standard is available on the Web at *www.osha-sic.gov/OshStd_data/1910.0119.html.* In addition to the PSM regulation itself, you will find appendices that list the chemicals covered by the regulation and nonmandatory compliance guidelines and recommendations for process safety management.

The **Hazard Communication (Hazcom) Standard** requires employers to identify hazardous chemicals in the workplace, label containers, obtain MSDSs, provide training, and prepare a written program for complying with the regulation. Non-industrial laboratory employees were covered by the Hazcom standard until 1991 when the Laboratory Standard was added to the regulations. In those circumstances where the Laboratory Standard applies, its specific regulations supersede similar provisions of the Hazcom Standard. Laboratories that are part of a production facility (generally quality control) remain covered by the Hazcom Standard and are not covered by the Laboratory Standard.

The **Laboratory Standard** was written to reflect laboratory conditions more closely. The Hazcom Standard pertains in many ways to large amounts of chemicals, such as the quantities needed in manufacturing. The Laboratory Standard is based on the concept of "laboratory scale," meaning that adjustments have been made to reflect the smaller quantities of chemicals used in laboratory settings. In 29 CFR 1910.1450, OSHA defines laboratory scale as "work with substances in which the containers used for reactions, transfers, and other handling of substances are designed to be easily and safely manipulated by one person. 'Laboratory scale' excludes those workplaces whose function is to produce commercial quantities of materials." The Laboratory Standard also requires that employers develop SOPs, ensure that fume hood and safety equipment are working properly, delineate the methods for determining and implementing control measures, identify circumstances requiring prior approval and persons responsible for implementing the plan, describe procedures for medical consultation and examination, provide training and information, and describe special precautions for dealing with particularly hazardous substances. The Hazcom Standard and Laboratory Standard are part of an employee's right to know about the chemical hazards in his or her workplace. The Laboratory Standard requires the appointment of a chemical hygiene officer and the preparation of a written CHP.

The **Process Safety Management (PSM) Standard** applies to facilities that process or handle specified quantities of listed (highly hazardous) chemicals or any flammable liquids and gases in quantities of 10,000 pounds or more. The regulation mandates specific safety planning and management procedures for preventing or minimizing the consequences of catastrophic releases of toxic, reactive, flammable, or explosive chemicals. A catastrophic release generally means a major release that threatens the lives or safety of the plant workers and neighboring community. The PSM Standard requires companies covered by the regulation to develop a comprehensive safety program that systematically identifies, evaluates, and prevents accidents resulting from failures in process, procedures, or equipment. In essence, PSM is simply rigorous, comprehensive, ongoing safety planning. Key elements of the PSM Standard include the following:

- Process Safety Information—Employers are required to compile written safety information on the technology, chemicals, and equipment used in the process. Documents may include, but are not limited to, MSDSs, process flow or block diagrams, and piping and instrument diagrams.

- Employee Involvement—Employers are required to develop a written plan of action regarding employee participation. The plan should provide details on how the company will consult with its employees on the conduct and development of process hazard analyses and other elements of the process safety management procedures.

- Process Hazard Analysis (PHA)—One of the most important elements of process safety management, the PHA is an organized, risk-based, systematic effort to identify and analyze the significance of potential hazards associated with the processing and handling of highly hazardous chemicals. Employers must complete a PHA for each process covered by the regulation. PHAs are discussed in more detail in the following section.

- Standard Operating Procedures (SOPs)—Companies must provide SOPs that contain clearly written instructions for the safe operation of all processes. The SOPs must include steps for each operating phase, describing operating limits, safety and health considerations, and safety systems. *You will learn more about SOPs in Section 8C.*

- Training—This aspect of the regulation mandates employee training covering specific safety and health hazards, emergency operations, and safe work practices.

- Incident Investigation—Employers must investigate within 48 hours all incidents that resulted in, or could have resulted in, catastrophic releases of the chemicals covered in the regulation.

- Emergency Planning and Response—Employers must develop and implement an emergency action plan.

- Compliance Audits—Employers must certify they have evaluated compliance with safety requirements every three years.

A basic familiarity with the PSM Standard is important because you may eventually work at a facility covered by the regulation. Furthermore, most industries and laboratories adopt, either voluntarily or through other regulations, many of the safety practices employed by the PSM Standard. The regulation also requires worker involvement in the safety planning process. Therefore, you are very likely to encounter some aspects of the PSM Standard early in your career.

Process Hazard Analysis as a Safety Planning Tool

The Process Hazard Analysis (PHA) forms the core of process safety management. The primary goal of a PHA is to recognize, predict, and assess the potential hazards associated with processes, materials, equipment, and procedures used in a laboratory or industrial facility. In this context, "hazard" is usually defined as some characteristic of the system that has potential to cause accidents or undesirable incidents. In some ways, the PHA methodology is like a complex (and very serious) game in which a team of safety experts, managers, and workers attempt to identify the potential for the occurrence and severity of accidents or incidents before they happen. PHA results are then used for planning corrective actions to remedy the identified safety deficiencies.

PHA safety experts use a variety of structured hazard evaluation techniques to evaluate in a thorough, systematic manner the potential for accidents to occur. Hazard evaluation techniques include the following:

- what-if questions,

- hazard and operability (HAZOP) analysis,

- failure mode and effects analysis (FMEA),

- fault- or event-tree analysis, and

- management oversight risk tree (MORT) analysis.

These methods range from wholly qualitative "what-if" questions to fully quantitative risk assessments. Most of the methods are rather complicated to apply because they involve statistical and specialized flow-chart analyses. Actual implementation is usually done by an experienced team of risk-assessment specialists. You can learn more about these methods by consulting *Guidelines for Hazard Evaluation Procedures, 2nd Edition with Worked Examples,* published by the American Institute of Chemical Engineers' Center for Chemical Process Safety (referenced in the Overview of this chapter).

Whatever the method used, all PHAs require some degree of risk assessment. According to *Guidelines for Incorporating Safety and Health into Engineering Curricula, Vol. 1: Laboratory Safety (Guidelines),* "risk is a measure of potential losses expressed as probability and magnitude." For a risk assessment, students or workers should ask four questions when preparing for a laboratory experiment or chemical production operation:

- What can happen to cause an accident or incident?

- What is the probability that these events will occur?

- What are the potential results of each of these events?

- How can each of these events be prevented or their effects at least be minimized?

A simplified version of a PHA (sometimes called a "preliminary hazard analysis") is readily adaptable to laboratory and bench-scale operations. According to *Guidelines,* an initial PHA begins with developing a list to describe the known and potential hazards associated with the following:

- all chemicals associated with the experiment, whether feeds, intermediates, products, or by-products;

- the experimental apparatus;

- any safety-related problems, such as the presence of incompatible chemicals, fire and explosion potential, and extremes of temperature and/or pressure;

- environmental factors, such as static electricity;

- equipment layout and assembly setup;

- operating procedures, including startup, operation/data collection, normal and emergency shutdown, maintenance, and cleanup; and

- safety equipment, such as pressure-relief valves and PPE.

Much of this information is obtainable from the MSDSs for the chemicals involved and from the operating instructions or manufacturers' brochures describing the equipment.

A sample PHA for an ammonia-based gas absorber is shown in the following table. The items in the left-hand column are generic and could apply to any system or subsystem chosen. In an industrial setting, the gas absorber might be one of many subsystems examined in a manufacturing process.

Sample PHA for Laboratory-Scale Gas Absorber Using an Air-Water-Ammonia System		
	PHA Step	Example Analysis
1.	System or subsystem	Gas absorber.
2.	Operating mode or condition	Normal operation.
3.	Hazardous element	Ammonia supply.
4.	Trigger event causing hazardous condition	Rupture of regulator diaphragm.
5.	Hazardous condition	Excessive ammonia supplied to the packed tower exceeds absorption capacity.
6.	Trigger event causing potential accident	Vent fan intended to blow clean gas discharge away from building not turned on.
7.	Potential accident or incident	Overhead "clean" gas stream dumps back into laboratory through general exhaust vent.
8.	Possible effects of accident	Overexposure of students or workers to ammonia.
9.	Possible preventative measures	Relocate absorber exhaust line to 6 feet above roof level on downwind side of prevailing winds, and require two people to verify that the fan is on and running before opening the ammonia main cylinder valve. Close main cylinder valve if regulator diaphragm ruptures.
Adapted from *Guidelines for Incorporating Safety and Health into Engineering Curricula,* Vol. 1. Used with permission.		

An experimental apparatus (the gas absorber) was selected for analysis. The experimenter then asked a series of "what-if" questions, such as "What if the regulator diaphragm ruptured and the vent fan was not on?" The possible effects were then analyzed and preventative measures selected. In practice, a PHA could involve generating many such tables for various subsystems, conditions, and scenarios. For this reason PHAs must be conducted by a team, with each member addressing areas of his or her particular expertise.

In the real world, enacting preventative measures for every possible scenario is impossible. For this reason, risk-assessment techniques are used to evaluate the probability versus the potential severity of an accident or incident. *This issue is discussed in Section 8B.*

Using SOPs to Build a Safety Checklist

Another important requirement of the PMS Standard (and other safety regulations) is the development of SOPs to cover equipment operation, chemical preparation, waste disposal, PPE use, and other important procedures. Most laboratories and industries incorporate and require the use of SOPs.

The nuclear accident described in the Laboratory Incident Reports in this chapter shows that not all SOPs are accurate, clearly written, or adequate in addressing safety issues. Existing SOPs may need continuous revision as equipment is updated or incident reports reveal new, unanticipated hazards. The student or worker should also be responsible for noting inaccuracies or safety deficiencies in SOPs and reporting them to the appropriate safety officer.

An SOP can be used as a starting point for a simple PHA. Each step in the SOP is examined for omitted or overlooked safety precautions. A common procedure is to ask a series of "what-if" questions regarding the possible results of malfunctions and human error at each step of the procedure, such as the following: "What if the electricity failed?" "What if a hose broke?" "What if the system overheated?" "What if someone spilled the chemical?" Even seemingly remote hazards should be considered and evaluated. The resulting list of potential hazards and the steps needed to mitigate them can then be incorporated into a revised SOP. *You will practice this process in Section 8C.*

Safety Meetings and Audits

Regular safety meetings in industry are mechanisms for reinforcing the mission of the company's chemical hygiene plan (CHP) and for fulfilling government Worker Right-to-Know laws. In your laboratory career, you will most likely participate in periodic safety meetings. Safety audits are a prudent means of ensuring that the laboratory environment is safe and unlikely to result in unexpected situations that can lead to accidents and injuries. A component of the audit involves inspecting the laboratory facilities and safety equipment for safe working conditions. Audits should be an integral part of the CHP and should be conducted as often as possible to ensure safe working conditions. *You will learn more about safety meetings and audits in Section 8D.*

Section 8B: Risk Assessment
Instructor Notes

As preparation for evaluating an actual SOP for safety, students perform a risk assessment on using a common piece of laboratory equipment—a Bunsen burner. The objective of the Exercise is for students to learn that step-by-step written instructions not only allow the same task to be done consistently by many different people, but can also serve as an outline for brainstorming potential hazards associated with a task.

Student Pages

- Section 8B: Risk Assessment—Background
- Section 8B: Risk Assessment—Exercise

Possible Playout of the Section

Students read the Background and then complete the Exercise in which they analyze each step of the instructions for potential risks and qualitatively assess the probability of that risk.

Section 8B: Risk Assessment

Background

Note: This section provides general recommendations for and information about risk assessment. However, you should always follow the procedures established by your organization's chemical hygiene plan.

We all know from personal experience and observation of events around us that some harmful events are more likely to occur than others and that some events cause more harm than others. In simple terms, risk is a measure of the likelihood that a potential hazard will cause an accident (or incident) versus the severity of the accident. Generally, the less likely or less harmful an event is, the smaller the amount of time and resources that are spent trying to minimize its effects. Accepted small risk associated with unlikely or marginally harmful events is called "assumed risk." As events carry more potential for harm and/or are more likely to occur, the risk increases. The higher the risk, the greater the amount of time and resources that are spent to prevent harmful events and/or minimize their impact. Risk assessment is the process of determining the risk associated with potential hazards so that appropriate prevention measures can be undertaken.

Risk assessors commonly use a semi-quantitative Risk Assessment Matrix to evaluate accident frequency and severity. A sample risk assessment matrix is shown below.

Hazard Frequency Levels / Hazard Severity Categories		Risk Assessment Matrix					
		Catastrophic 5	Fatal 4	Critical 3	Marginal 2	Minor 1	Negligible 0
Frequent	5	A	A	A	A	A	B
Probable	4	A	A	A	B	B	B
Occasional	3	A	A	A	B	B	B
Remote	2	A	B	B	C	C	C
Improbable	1	B	C	C	D	D	D
Incredible	0	C	D	D	E	E	E

The left column provides a listing of the hazard frequency levels. In a detailed risk assessment, frequency categories would be assigned a probability based on actual statistical data. Probabilities are usually measured as frequency of occurrence over a given length of time (for example, number of occurrences per year).

Hazard severity categories (shown across the top of the Risk Assessment Matrix) provide a qualitative measurement of the worst outcome resulting from failures, malfunctions, or human error. The levels and their exact definitions will vary with each risk assessment, depending on the system or operations being analyzed. The sample Risk Assessment Matrix uses the following definitions:

- Catastrophic—Results in injuries or fatalities not only to coworkers but to the neighboring community

- Fatal—Results in death to operator and possible injuries to coworkers

- Critical—Results in serious or life-threatening injury to operator and/or serious equipment breakdown

- Marginal—Results in injury to a worker that requires some medical treatment

- Minor—Results in minor injury, requiring minimal or no treatment

- Negligible—At most results in psychological distress to worker(s)

In the sample Risk Assessment Matrix, the hazard information collected from the site was converted to a risk index by using an A–E grading system with the following categories:

- A—Unacceptable risk. Changes in safety procedures or system design are required.

- B—Moderate risk. Changes in safety procedure may be required.

- C—Marginal risk. Adopt changes if practical and cost-effective.

- D—Assumed risk/negligible risk. Adopt changes if minor and/or cost-effective.

- E—Assumed risk. No changes in procedure or design required.

In general, the higher the hazard risk index, the greater should be the effort of safety planning to reduce or eliminate the hazard. The level at which preventive measures are undertaken depends on the system or hazards involved. For example, the first Apollo astronauts were required to wear biological isolation garments after leaving their capsule and had to stay in quarantine for a period of time, in case they had brought back harmful microbes from the surface of the moon. Based on our knowledge of the moon even in 1969, this possibility was improbable or even incredible, but the potential consequences of such an event could have been catastrophic for the entire Earth's population. Therefore, it was reasonable to enact modest safeguards. In contrast, an isolated meteorite impact on a nuclear power plant is also incredible and possibly catastrophic, but the safeguards necessary to prevent this hazard—short of banning nuclear power altogether—are simply not cost-effective or possible to implement. Depending on the operation, management and critical personnel may determine that levels of E, D, and in some cases C are simply part of "assumed risk." A hypothetical area of assumed (or negligible) risk is indicated by the shaded area in the Risk Assessment Matrix.

Section 8B: Risk Assessment
Exercise

In this Exercise, you will develop a step-by-step written procedure for using a Bunsen burner and assess the potential risks involved in each step of the procedure. The objective of this Exercise is to help you understand how a set of written instructions can be used as an outline for safety planning and hazard identification.

To conduct a risk assessment on the use of a Bunsen burner, begin by writing the procedural steps one should follow to safely use the burner. (Start with a burner, a hose, a gas source, an ignition source, and other materials that are needed. Assume nothing about the previous use or condition of these items.) Although this task may seem trivial, you should assume you are preparing this procedure for an audience that has never used a Bunsen burner and is unfamiliar with basic laboratory procedure and safety practices.

Once the procedure is written, brainstorm a list of possible hazards associated with the activity. Assign each hazard a frequency score from 0 to 5 and a severity score from 0 to 5. Then assign a hazard risk index for each step.

Review and compare your results with the class or groups assigned by the instructor.

Section 8C: Evaluating an SOP
Instructor Notes

The objective of this section is for students to learn about SOPs. They read and analyze an existing SOP to determine whether it includes enough safety procedures, applying all the other specific procedures they have learned from this book.

Student Pages

- Section 8C: Evaluating an SOP—Background
- Section 8C: Evaluating an SOP—Exercise

Possible Playout of the Section

Students read about evaluating SOPs in the Background. In the Exercise they read an SOP and consider whether it includes adequate safety information. Their evaluation can be done either as a discussion, in small groups, or as a written assignment.

You may wish to follow up this Exercise by allowing students to develop an SOP (including all relevant safety information) for a laboratory procedure they are comfortable with. You may also allow the students to work in teams to model an industry approach to writing SOPs.

Section 8C: Evaluating an SOP

Background

Note: This section provides general recommendations for and information about SOPs. However, you should always follow the procedures established by your organization's chemical hygiene plan.

A standard operating procedure (SOP) is a written procedure that must be followed when performing a laboratory task. The SOP lists all the steps that should be taken in that particular procedure. You may not have encountered documents that were actually called SOPs, but you have probably used recipes, instructions for home appliances, and laboratory manuals, which are all really SOPs. OSHA's Laboratory Standard requires that a chemical hygiene plan include SOPs and that those SOPs include health and safety information. Because of the potentially hazardous nature of procedures carried out in the laboratory, SOPs must be well written and clearly understood.

In industry, SOPs are usually written by a team, including a worker who routinely performs the task to be explained, a supervisor, and a safety coordinator. In the workplace, you may be part of a team that develops SOPs. Writing an SOP involves keeping the audience and purpose in mind. The audience is someone who will be performing the task being described. You should consider what this person will need to know in order to perform the procedure correctly.

SOPs should include the following:

- identifying information such as a title or number,
- the date it will go into effect and any revision dates,
- the author's name and contact information,
- the task description and where it should be done,
- the number of people required,
- the required skill level of the workers,
- equipment and supplies required,
- PPE and emergency equipment required, and
- the finished product or objective.

All terms, acronyms, and regulations referred to in the SOP should be defined. All training necessary to perform the operation described should be explained.

Although SOPs are required to have health and safety information, people have different philosophies about whether the SOP should be all-inclusive or whether it should reference external documents for details. For example, an SOP involving the use of concentrated sulfuric acid might not include routine safety considerations if the audience was assumed to be an experienced chemist or technician. The writer of the SOP might assume the reader would obtain the necessary information (such as an MSDS) and act on it appropriately. In another case, an SOP for a process requiring a particular solution (such as a standardized acid solution) might refer to a different SOP for making the solution rather than embedding the solution-making instructions.

The SOPs you encounter in the workplace are not guaranteed to provide all the relevant information you need to do the procedure safely. For your own safety and the safety of your colleagues, it will be your responsibility to find additional safety information when the SOP does not provide it.

Section 8C: Evaluating an SOP
Exercise

In this Exercise you read and analyze an existing SOP (developed by the polymer industry and reproduced below) to determine whether it includes sufficient safety procedures.

Use your knowledge of safety procedures to identify the statements in this SOP that pertain to safety. Assume that the reader of this SOP is unfamiliar with the chemicals used in the procedure. Does the SOP include **everything** that the reader needs to know before beginning the procedure? If not, explain what is missing. Using appropriate reference materials (such as MSDSs), add all necessary safety precautions to this SOP.

DETERMINATION OF TRACES OF PEROXIDES IN ORGANIC SOLVENTS
SPECTROPHOTOMETRIC METHOD

Key Words: Peroxides, Solvents, Spectrophotometric, Trace, Visible

I. **PRINCIPLE**

A sample is dissolved in acetic acid-chloroform solvent and de-aerated with nitrogen. Potassium iodide reagent is then added and the sample is allowed to react in the dark under a nitrogen blanket. An equivalent amount of iodine is released which is measured spectrophotometrically and related to the original peroxide content of the sample.

II. **SCORE**

The method is applicable to samples containing 2 ppm or more of active oxygen. Numerous classes of organic solvents can be analyzed including alcohols, acids, esters, ketones, alkanes, olefins, and conjugated olefins. Solids can also be analyzed provided they are soluble in the acetic acid-chloroform medium. Compounds which have been successfully analyzed include benzene, chloroform, isopropanol, methanol, pentane, hexane, toluene, diethyl ether, acetone, vinyl acetate, ethyl acetate, octenes, heptenes, cyclohexene, isoprene, acrolein, tetralin, and sorbic acid. Interference will be encountered if other oxidizing substances are present in the sample. Refractory peroxides such as di-*tert*-butyl peroxide and dicumyl peroxide cannot be determined by this method.

III. **SAFETY PRECAUTIONS**

Chloroform is an oral and systemic poison with well-known anesthetic properties. It is irritating to the skin, eyes, and mucous membranes and is a suspected carcinogen. The threshold limit value (TWA, ACGIH) is 10 ppm or 50 mg/m^3. Avoid prolonged contact and use only in a hood.

IV. **APPARATUS AND EQUIPMENT**

1. Spectrophotometer, Perkin-Elmer Model Lambda 3B or equivalent with matched 1 cm borosilicate cells and covers.

V. **REAGENTS**

1. Acetic Acid-Chloroform Solvent, 2 to 1: Add 300 mL of chloroform to 600 mL of acetic acid and mix well.

2. Potassium Iodide, 50%: Dissolve 20 g of KI in 20 mL of deaerated water. Prepare this reagent just prior to use.

3. Standard Iodine Solution, 80 µg Active Oxygen per mL: Dissolve 0.1270 g of pure iodine in about 50 mL of 2:1 acetic acid-chloroform solvent. Transfer the solution to a 100 mL volumetric flask, dilute to volume with a 2:1 acetic acid-chloroform, and mix well. This solution contains 1.27 mg of iodine per mL which is equivalent to 80.0 µg of active oxygen per mL.

VI.   REQUIREMENTS

The maximum sample size required for this demonstration is 15 mL. An elapsed time of approximately 2.5 hours is required to complete the analysis. The analyst hours required to complete one and six determinations are approximately 1.5 and 2.5 hours respectively.

VII.   PROCEDURE

A. Preparation of Standard Calibration Curve

1. Pipet 0, 1, 2, 3, 4, and 5.00 mL aliquots of standard iodine (80.0 µg active oxygen per mL solution) into 25 mL volumetric flasks and dilute to volume with 2:1 acetic acid-chloroform solvent.

2. Purge each solution with a fine stream of nitrogen for 1 min by means of a hypodermic needle or Pasteur pipet with the tip inserted to the bottom of the flask.

3. Pipet 1.0 mL of freshly prepared 50% KI into each flask and continue the nitrogen purge for 1 minute. Stopper and mix thoroughly.

4. Immediately measure the absorbance of each solution at 470 nm using covered 1 cm borosilicate cells and water as a reference.

5. Subtract the absorbance of the blank from the absorbance values of the standards and plot absorbance vs. micrograms of active oxygen per 25 mL. This curve covers the range 0–400 mg of active oxygen.

B. Analysis of Samples

1. Transfer a sample containing up to 400 µg of active oxygen to a previously tared 25-mL volumetric flask. Restopper and again weigh to the nearest 0.1 mg to obtain the exact sample weight. See Note 1.

2. Dilute to volume with 2:1 acetic acid-chloroform and mix thoroughly.

3. Remove the stopper and purge the solution with a fine stream of nitrogen for 1½ minutes by means of a hypodermic needle or Pasteur pipet inserted to the bottom of the flask.

4. Pipet 1.0 mL of freshly prepared 50% KI solution into the flask and continue the nitrogen purge for 1 minute.

5. Discontinue the nitrogen flow and immediately stopper the flask and mix thoroughly. Allow the flask to stand in the dark for 1 hour. See Note 2.

6. Measure the absorbance of the solution at 470 nm using covered 1 cm borosilicate cells and water as a reference. Absorbance measurements should be made as rapidly as possible to minimize the effect of air oxidation.

7. Subtract the absorbance of a blank carried through the entire procedure from the absorbance of the sample and obtain the micrograms of active oxygen present in the sample by reference to the calibration curve. See Note 3.

8. Calculate the ppm of active oxygen present in the sample.

VIII.   CALCULATIONS

$$\text{Active Oxygen (ppm)} = \frac{\text{Active Oxygen Found (µg)}}{\text{Sample Weight (g)}}$$

See Note 4

IX. REPORT

Report the active oxygen content to the nearest ppm.

X. PRECISION AND ACCURACY

Precision and accuracy data are not available.

XI. NOTES

1. Samples up to 15 mL may be taken for analysis provided they are completely miscible when diluted to 25 mL with acetic acid-chloroform solvent and remain miscible after addition of KI. Some precipitation of KI may occur on standing but this does not affect the analysis. Samples which are very low in active oxygen or are only slightly miscible with the acetic acid-chloroform solvent can be analyzed using a special cell which covers the range of 0 to 40 µg active oxygen. Sensitivity is increased by making absorbance measurements at 410 nm. Details of this cell and its use are given in Reference 2.

2. While reaction with most peroxides is quantitative within a few minutes, a longer reaction time is required for the more stable peroxides. A general reaction time of one hour is therefore specified in the procedure.

3. If the sample solution is colored when diluted to volume with acetic acid-chloroform solvent, an appropriate correction should be made as follows: Transfer an identical weight of sample to a 25-mL volumetric flask and dilute to volume 2:1 acetic acid-chloroform. Add 1 mL of water and mix thoroughly. Measure the absorbance of the solution at 470 nm in 1 cm cells against water as a reference. Subtract the absorbance of this solution from the net absorbance value of the sample as obtained in steps 6 and 7 under Analysis of Samples.

4. If a specific peroxide is known to be present, the ppm of active oxygen can be converted to ppm of peroxide by multiplying by (100/F) when F = the percent of active oxygen present in the pure peroxide.

XII. REFERENCES

1. Heaton, F.W.; Vir, N. J. Sci. Food Agri. 1958, 9, 781.

2. Banerjee, D. K.; Budke, C. C. Anal. Chem. 1964, 36, 792–796.

3. Banerjee, D. K.; Budke, C. C. Anal. Chem, 1964, 36, 2367–2368.

Section 8D: Safety Meetings and Audits
Instructor Notes

The objective of this section is to help students understand the role of safety meetings and audits in laboratory safety planning.

Student Pages

• Section 8D: Safety Meetings and Audits—Background

Possible Playout of the Section

Students read the Background and discuss the role of safety meetings and audits in laboratory safety planning.

You may wish to conduct safety meetings with students to model the safety meetings they are likely to encounter on the job and to reinforce the safety ideas being taught concurrently in the laboratory. Students can take turns being responsible for safety meetings by leading discussions and selecting topics. Additionally, you may wish to have students design and implement laboratory safety audits/inspections as part of the safety meeting. **Chapter 5, Section B: Working with Electrical Laboratory Equipment** includes an electrical safety inspection. Students could use this as a model for developing other audits/inspections.

Section 8D: Safety Meetings and Audits

Background

Note: This section provides general recommendations for and information about safety meetings and audits. However, you should always follow the procedures established by your organization's chemical hygiene plan.

Safety meetings and audits play an important role in laboratory safety planning. In your laboratory career, you will most likely participate in both.

Safety Meetings

Regular safety meetings in industry are mechanisms for reinforcing the mission of the company's chemical hygiene plan (CHP) and for fulfilling government Worker Right-to-Know laws. In your laboratory career, you will most likely participate in periodic safety meetings conducted by a person designated as the safety officer or chemical hygiene officer (CHO). The frequency, duration, and organization of the safety meetings are determined in writing in the CHP. Safety meetings are commonly held once a month on a designated day and time and allow communication of safety information and new regulations, reinforcement of safe practices, and sharing of concerns. CHPs commonly mandate worker attendance at the safety meetings. Government regulations for industry recommend that attendance sheets, minutes, and copies of handouts for safety meetings be kept.

The Occupational Safety and Health Administration's (OSHA) regulations, as found in 29 CFR 1910, require employers to provide appropriate training for personnel working with hazardous chemicals. The training must be related to relevant workplace risks and must be consistent with the background of the employees. Safety meetings are often a mechanism for providing this training. Topics typically include the following:

- material safety data sheets (MSDSs) for new chemicals in the laboratory,
- new hazard potentials in the laboratory,
- review of hazardous materials standard operating procedures (SOPs),
- personal protective equipment (PPE) refresher courses,
- misuse of medications, alcohol, and illicit drugs on the job,
- CPR and first aid training,
- machine and equipment safety,
- avoiding back injuries,
- falling objects,
- stairway safety/slipping and falling,
- emergency response training, and
- prior approval reviews.

In addition to providing the safety training just discussed, safety meetings often include the topics discussed in the following paragraphs.

Incident Reporting

At safety meetings, the safety officer or laboratory supervisor typically reports briefly on laboratory accidents that have occurred since the last safety meeting. The reports are not presented with the intent of embarrassing a particular individual. Rather, they serve as a reminder that safety must be foremost on everyone's mind and to alert laboratory workers to be more cautious about activities associated with the accident. Reviewing laboratory accidents may also lead to updated safety procedures. Safety programs in industry are constantly being updated as new information is developed or discovered.

Feedback About Work Conditions Related to Safety

Safety meetings can be important opportunities for laboratory workers to communicate concerns or needs related to current safety practices. Many safety officers use this time as a forum to gain valuable input about how current safety practices are working. Issues such as discomfort or problems using certain PPE can be brought out and resolved. In many situations, PPE is improperly used or not used at all because of problems encountered with it. Feedback about the effectiveness of the CHP is also an important issue raised at safety meetings. Informal and surveyed information from safety meetings can become part of the process of evaluating and updating the CHP.

Off-the-Job Safety

Safety meetings can cover off-the-job safety as well. Appropriate topics for safety meetings include safety in the home and personal health issues. Part of maintaining a safe work environment is instilling a safety attitude that is practiced at home and in the community. Days off for injury are most likely to result from incidents and accidents occurring away from the workplace.

Conducting a Laboratory Audit

Safety audits can ensure that the laboratory environment is safe and unlikely to result in unexpected situations that can lead to accidents and injuries. A component of the audit involves inspecting the laboratory facilities and safety equipment for safe working conditions. Audits should be an integral part of the CHP and should be conducted as often as possible to ensure safe working conditions.

The laboratory safety audit, as with other practices in the laboratory, requires a written plan detailing the safety conditions and practices that will be assessed and the methods by which they will be assessed. Safety audits should be designed to fit the specific needs of particular laboratory facilities, PPE, or safety equipment.

The purposes of a safety audit are prevention of accidents and preparation for handling an accident should one occur. The occurrence of an accident implies the existence of unknown or uncontrolled hazards in the laboratory. In *Improving Safety in the Chemical Laboratory*, J. A. Young, a chemical safety and health consultant, classified accidents and hazards in the laboratory as follows:

- "struck by" exposures, where an object (an item falling off a shelf, broken glass from an explosion or implosion, something thrown) hits a person;

- "strike against" exposures, where someone in motion collides with a fixed object;

- "caught on" exposures, where someone's hair or clothing gets caught on something in the laboratory;

- "caught between" exposures, where a person is stuck between two objects;

- overexertion; and

- chemical exposures, including inhalation of toxic gases, dusts, or mists; irritating or corrosive substances in contact with mucous membranes or skin; temperature extremes, such as extreme cold from cryogenic materials or intense heat from fire, uncontrolled reactions, or very hot substances; sources of radiation; and deficiency of oxygen in an enclosed space.

The purpose of safety audits is to identify unsafe conditions such as the following listed in *Improving Safety in the Chemical Laboratory:*

- inoperative emergency and safety equipment;

- insufficient training in the use of emergency equipment;

- poor housekeeping;

- narrow clearances between passageways, between laboratory benches, or near exit doorways;

- inadequate or unsafe shelving and storage;

- insufficient illumination;

- crowded laboratory bench surfaces;

- improper electrical wiring;

- inoperative warning systems;

- lack of machine guards;

- noncompliance with safety procedures;

- unsafe chemical storage; and

- poorly maintained mechanical equipment.

Laboratory Work Space

Work benches and the areas around the benches should be inspected for cleanliness and proper chemical storage. Chemicals in the area should be labeled appropriately and be placed where they do not cause dangerous clutter. Floor spaces around the laboratory benches, hoods, aisles, and laboratory equipment should be free of obstacles that could cause falls. Exits and emergency escape windows should not be blocked with objects that could hinder access. Chemical spills should be cleaned from work areas and laboratory equipment.

Personal Protective Equipment

Regular inspection should be done to monitor whether workers or students in the laboratory are using the proper PPE. Proper gloves, goggles, and laboratory apparel should be in use when chemicals are being handled. PPE must be used properly to prevent contamination of the surroundings and should be removed and washed when completing a task or leaving the laboratory to reduce the risk of contaminating non-chemical areas. The inspection should also include whether PPEs are being worn out of the laboratory to areas not designated for chemicals.

Safety Equipment

Equipment such as laboratory hoods should show current inspection for flow rate and must be in use when appropriate laboratory procedures are being carried out. Emergency equipment such as eye washes, fire blankets, fire extinguishers, and safety showers also need to be inspected regularly. Inspection tags provided by the maintenance, housekeeping, or safety department should be checked. Sign-off forms can be designed for equipment not inspected regularly by these departments. Safety equipment should not be blocked by any objects that could hinder accessibility. Lastly, a sign clearly identifying the location of safety equipment should be placed in an easily visible location.

Documenting a Safety Audit

Checklists appropriate for the audit can be designed for the particular laboratory. A free laboratory safety checklist is available at the Laboratory Safety Institute's website, *www.Labsafety.org.* Dates, times, and personnel involved in an audit should be documented and witnessed. Monthly or quarterly summary sheets can be designed to analyze safety trends for reporting at safety meetings or rating the effectiveness of the CHP. Communicating the results of laboratory audits serves as a learning tool for recognizing the need to maintain safe working conditions in the laboratory. Safety audit reports can be incorporated into safety meetings.

Glossary

ACGIH: American Conference of Governmental Industrial Hygienists. An organization of professionals in educational institutions and/or governmental agencies engaged in occupational safety and health programs. ACGIH develops and publishes recommended occupational exposure limits for chemical substances and physical agents.

Acute: Immediate adverse effect, generally occurring in a day or less.

ANSI: American National Standards Institute. A privately funded organization that identifies industrial/public national consensus standards and coordinates their development. Many ANSI standards relate safe design/performance of equipment and safe practices or procedures. Some of these standards have been adopted as federal laws.

Assumed risk: Accepted small risk associated with unlikely or marginally harmful events.

Bonding: Connecting two items by a wire so that no difference in electric potential can exist between them.

Carcinogen: A substance or material that causes cancer.

CAS Registry Number: A number designating a particular chemical; this number is assigned by the Chemical Abstract Service, an organization that indexes information published in the American Chemical Society's *Chemical Abstracts.* CAS numbers have no chemical significance; they indicate where information about a given substance can be found in *Chemical Abstracts.*

CBC: Critical behavior checklist. Used in safety coaching.

CERCLA: The Comprehensive Environmental Response, Compensation, and Liability Act. Commonly referred to as Superfund, CERCLA is found at Public Law PL 96–510 and 40 CFR 300. CERCLA brought active government involvement to emergency response, site remediation, and spill prevention. CERCLA provides for identification and cleanup of hazardous materials released on the land and into the air, waterways, and groundwater. It covers areas affected by newly released materials and older leaking or abandoned dump sites. CERCLA established the Superfund, a trust fund to help pay for cleanup costs of hazardous materials sites. See also **SARA.**

CFR: Code of Federal Regulations. A collection of regulations established by federal law.

CHP: Chemical hygiene plan; found in 29 CFR 1910.1450, OSHA's Laboratory Standard. By law, this written plan must include specific work practices, standard operating procedures, equipment, engineering controls, and policies to minimize employees' exposure to potentially hazardous chemicals used in their work area. This OSHA standard provides for training, employee access to information, medical consultations and examinations, hazard identification procedures, respirator use, and record-keeping practices.

Chronic: Acting over a long period of time (longer than one day) to cause an adverse effect.

Clean Air Act: Found at 40 CFR 50–69. A national air quality act that, among other provisions, was written to improve the nation's air quality and to protect the health and welfare of citizens from being endangered by air pollution.

Clean Water Act: Found at 40 CFR 100–149 and 40 CFR 400–469. A national air quality act that, among other provisions, was written to improve the nation's water quality and to protect the health and welfare of citizens from being endangered by water pollution. Also known as the Federal Water Pollution Control Act.

Combustible: Liquids with a flash point at or above 100°F (29 CFR 1910.1200).

Corrosive: Causes visible destruction or irreversible alterations of living tissue through chemical action at the site of contact (OSHA 29 CFR 1910.1200).

Dose-response relationship: A function that describes the effects of exposure on living organisms to a chemical versus the dosage of that chemical.

DOT: U.S. Department of Transportation. A federal agency that regulates the transportation of materials, protecting the public as well as law enforcement, fire, and other emergency-response personnel. DOT classifications specify the use of appropriate warnings, such as Oxidizing Agent or Flammable Liquid.

EPA: U.S. Environmental Protection Agency. A federal agency with environmental protection, regulatory, and enforcement authority.

Explosive: A substance that produces a sudden, almost instantaneous release of pressure, gas, and heat when subjected to sudden shock, pressure, or high temperature.

Flammable: A substance whose flash point is 100°F or less.

Flammability range: The range of concentrations of a flammable gas or vapor in air between which ignition can occur.

Flash point: The minimum temperature at which a liquid or solid gives off vapor sufficient to form an ignitable fuel-air mixture.

FMEA: Failure mode and effects analysis. A suggested methodology, listed in 29 CFR 1910.119, that employers can use to evaluate hazards of a process, required by the PSM Standard.

Grounding: Connecting an item or group of items and the ground together with a wire so no electric potential can exist between the item(s) and the ground.

Hazard: A source of danger.

Hazard Diamond (NFPA Hazard Rating): The National Fire Protection Agency (NFPA) visual rating system that addresses the severity of health, flammability, reactivity, and related hazards that a material may pose during a short-term acute exposure caused by a fire, spill, or similar emergency, as listed in NFPA 704.

Hazcom (Hazard Communication) Standard: Section of the OSHA regulations (found in 29 CFR 1910.1200) that requires employers to identify hazardous chemicals in the workplace, label containers, obtain material safety data sheets (MSDSs), provide training, and prepare a written program for complying with the regulation.

HAZOP Analysis: Hazard and operability analysis. A suggested methodology, listed in 29 CFR 1910.119, that employers can use to evaluate hazards of a process, required by the PSM Standard.

HMIS: The hazardous materials identification system developed by the National Paint and Coatings Association (NPCA) to provide information on the acute health, reactivity, and flammability hazards encountered in the workplace. This system includes temperatures under fire conditions (especially for flammability and reactivity). A number is assigned to a material indicating degree of hazard, from 0 for the least up to 4 for the most severe. Letters designate personal protective equipment.

IDLH: Immediately dangerous to life and health. The maximum concentration of a potentially hazardous substance one could be exposed to during a 30-minute period and escape without experiencing irreversible health effects or impairing symptoms. This number is used in respirator selection. Carcinogenic effects are not considered when setting these values.

Irritant: Causes a reversible inflammatory effect on living tissue by chemical action at the site of contact but is not corrosive (OSHA 29 CFR 1910.1200).

Laboratory Standard: Section of the OSHA regulations (found in 29 CFR 1910.1450). It is similar to the Hazcom Standard but applies to laboratory-scale quantities of chemicals. It requires employers to develop standard operating procedures (SOPs), ensure that fume hoods and safety equipment are working properly, delineate the methods for determining and implementing control measures, identify circumstances requiring prior approval and persons responsible for implementing the plan, describe procedures for medical consultation and examination, provide training and information, and describe special precautions for dealing with particularly hazardous substances.

LC_{50}: Lethal concentration 50, median lethal concentration. The concentration of a material, generally in air, that on the basis of laboratory tests is expected to kill 50% of a group of test animals when administered as a single exposure in a specific time period. Generally this value is based on the inhalation of a substance over a 1-hour period. An LC_{50} is generally expressed as parts of material per million parts of air, by volume (ppm) for gases and vapors, as micrograms of material per liter of air (μg/L), or milligrams of material per cubic meter of air (mg/m^3) for dusts and mists, as well as for gases and vapors.

LCSS: Laboratory Chemical Safety Summaries. A chemical safety information document format developed specifically to meet the needs of laboratory workers by the National Research Council due to the limitations of MSDSs; they do not replace MSDSs but are intended to supplement them.

LD$_{50}$: Lethal dose 50. The single dose of a substance that causes the death of 50% of an animal population from exposure to the substance. Generally this value is based on an exposure route other than inhalation, such as oral or intravenous. An LD$_{50}$ is usually expressed as milligrams or grams of material per kilogram of animal weight. (mg/kg or g/kg). The animal species, means of administration (oral, intravenous, etc.), and dose are usually stated.

LEL: Lower explosive limit. Refers to the lowest concentration of a combustible or flammable substance (gas, vapor, or dust), by percent volume in air, that produces an explosion or that ignites when it contacts an ignition source (high heat, electric arc, spark, or flame). Any concentration below the LEL in air is not rich enough to be ignited.

LFL: Lower flammability limit. Refers to the lowest concentration of a combustible or flammable substance (gas, vapor, or dust), by percent volume in air, that produces a fire or that ignites when it contacts an ignition source (high heat, electric arc, spark, or flame). Any concentration below the LFL in air is not rich enough to be ignited.

MSDS: Material safety data sheet. A fact sheet summarizing information about material identification; hazardous ingredients; health, physical, and fire hazards; first aid; chemical reactivities and incompatibilities; spill, leak, and disposal procedures; and protective measures required for safe handling and storage. OSHA requires MSDSs to be made available by those who make, distribute, and use hazardous materials so that the hazards associated with those materials can be effectively communicated.

MORT analysis: Management oversight risk tree analysis. A methodology employers can use to evaluate hazards of a process. MORT analysis is considered to be an appropriate equivalent to the methodologies listed in the PSM Standard.

Mutagen: A substance or material that causes the genetic make-up in a cell to change.

NFPA: National Fire Protection Association. An international voluntary membership organization formed to promote and improve fire prevention and protection as well as to establish safeguards against loss of life and property by fire. Best known for the National Fire Code, recommended practices, and manuals developed (and periodically updated) by NFPA committees.

NIOSH: National Institute of Occupational Safety and Health. The agency of the Public Health Service that tests and certifies respiratory and air-sampling devices. NIOSH recommends exposure limits to OSHA for substances, researches occupational safety, and investigates incidents.

OSHA: Occupational Safety and Health Administration. The regulatory and enforcement agency for safety and health in most U.S. industrial sectors, it is part of the U.S. Department of Labor.

OSH Act: The Occupational Safety and Health Act of 1970, found at 29 CFR 1910, 1915, 1918, and 1926. OSHA has jurisdiction. This regulation ensures the safety and health of workers in firms larger than 10 employees. Its goal is to set standards of safety that prevent injury and illness among the workers. Minimizing employee exposure and informing employees of the dangers of materials are key factors.

PEL: Permissible exposure limit. These limits are established by OSHA and have the force of law. A PEL may be expressed as a time-weighted average (TWA) limit, a short-term exposure limit (STEL), or as a ceiling exposure limit. A ceiling limit must never be exceeded instantaneously even if the TWA exposure limit is not violated.

PHA: Process hazard analysis. An organized, risk-based, systematic effort to identify and analyze the significance of potential hazards associated with the processing and handling of highly hazardous chemicals, required by the PSM Standard.

Poison: A substance that causes the disturbance, disease, or death of a living organism.

PPE: Personal protective equipment. Devices or clothing worn to help protect a worker from direct exposure to physical and chemical hazards. Examples include gloves, respirators, safety glasses, and ear plugs.

PPM: Parts per million.

PSM Standard: Process Safety Management Standard. Section of the OSHA regulations that applies to facilities that process or handle specified quantities of listed (highly hazardous) chemicals or any flammable liquids and gases in quantities of 10,000 pounds or more.

RCRA: Resource Conservation and Recovery Act. Found at 40 CFR 240–271. EPA has jurisdiction. Its purpose is to promote the proper disposal of hazardous and nonhazardous waste, as well as the recovery, recycling, and reuse of waste. RCRA's major emphasis is the control of hazardous waste disposal.

Risk assessment: The process of determining the risk associated with potential hazards so that appropriate prevention measures can be undertaken.

SARA: Superfund Amendments and Reauthorization Act. This federal act is a reauthorization of CERCLA. Title III of SARA, the Emergency Planning and Community Right-to-Know Act is the first piece of legislation to provide the general public access to information about the chemicals used, stored, and emitted in their communities. See also **CERCLA.**

Sensitizer: After repeated exposure causes a substantial proportion of people or animals to have an allergic reaction in normal tissue (OSHA 29 CFR 1910.1200).

SOP: Standard operating procedure. Contains clearly written instructions for the safe operation of a process. An SOP must include steps for each operating phase, as well as describing operating limits, safety and health considerations, and safety systems.

Teratogen: A substance or material that disrupts fetal development, often seen as birth defects, but does not necessarily change the genetic make-up of cells.

Threshold dose: The concentration below which a chemical is not considered to be harmful.

TLV: Threshold limit value. A term ACGIH uses to express the maximum airborne concentration of a material to which most workers can be exposed during a normal daily and weekly work schedule without adverse effects. A "worker" is a healthy individual; a "healthy" worker is defined as a 150-pound male, age 25–44. The young, old, ill, or naturally susceptible have lower tolerances and need to take additional precautions. TLVs are recommended exposure limits that OSHA may or may not enact into law.

TLV-STEL: Threshold limit value-short term exposure limit. The maximum airborne concentration of a material to which most workers can be exposed during a continuous exposure period of 15 minutes, with a maximum of four such periods per day, with at least 60 minutes between exposure periods, and provided the daily TLV-TWA is not exceeded.

TLV-TWA: Threshold limit value-time weighted average. The maximum allowable airborne concentration of a material to which most workers can be exposed during for a normal 8-hour workday or 40-hour week.

Toxicology: The study of substances that are harmful to living organisms.

Toxin: A poisonous or toxic substance secreted by a living organism.

TSCA: Toxic Substances Control Act (found in 40 CFR 701–763). TSCA establishes complex reporting requirements for any facility that manufactures, processes, or imports regulated amounts of toxic substances.

UEL: Upper Explosive Limit. The highest concentration of a combustible or flammable substance (gas, vapor, or dust), by percent volume in air, that produces an explosion or that ignites when it contacts an ignition source (high heat, electric arc, spark, or flame). Any concentration above the UEL in air is too rich to be ignited.

UFL: Upper Flammability Limit. The highest concentration of a combustible or flammable substance (gas, vapor, or dust), by percent volume in air, that produces a fire or that ignites when it contacts an ignition source (high heat, electric arc, spark, or flame). Any concentration above the UFL in air is too rich to be ignited.

References

The ABC's of the EPA; CERTIFIED: Weymouth, MA.

Hazard Communication Standard. *Code of Federal Regulations;* 29 CFR 1910.1200, 1999. (This is available free online at www.osha.gov.)

Kubasek, N.K.; Silverman, G.S. *Environmental Law;* Prentice Hall: Englewood Cliffs, NJ, 1994.

The MSDS Pocket Dictionary, 2nd ed.; Genium: New York, 1996.

Occupational Exposure to Hazardous Chemicals in Laboratories Standard. *Code of Federal Regulations;* 29 CFR 1910.1450, 1999. (This is available free online at www.osha.gov.)

Pocket Guide to MSDSs and Labels: Your Keys to Chemical Safety; Business and Legal Reports: Madison, CT, 1991.

Process Safety Management of Highly Hazardous Chemicals Standard. *Code of Federal Regulations;* 29 CFR 1910.119, 1999. (This is available free online at www.osha.gov.)

Prudent Practices in the Laboratory: Handling and Disposal of Chemicals; National Academy: Washington, DC, 1995.